Buffalo Creek Chronicles

Buffalo Creek Chronicles

Diary of a Cattle Ranch on the Southern Plains

Gary Lantz

Don House

WITH REFLECTIONS ON RANCH LIFE BY

Sue Selman

PHOENIX INTERNATIONAL, INC.
FAYETTEVILLE
2002

Printed in China

05 04 03 02 4 3 2 1

Designed by John Coghlan

Inquiries should be addressed to:

Phoenix International, Inc.
1501 Stubblefield Road
Fayetteville, Arkansas 72703
Phone (479) 521-2204
www.phoenixbase.com

Library of Congress Cataloging-in-Publication Data

Lantz, Gary, 1947–
Buffalo Creek chronicles : diary of a cattle ranch on the southern Plains / Gary Lantz ; Don House ; with reflections on ranch life by Sue Selman.
p. cm.
ISBN 0-9650485-9-4
1. Ranch life—Oklahoma. 2. Ranch life—Great Plains. 3. Selman, Sue, 1948—Childhood and youth. 4. Ranchers—Oklahoma—Biography. 5. Lantz, Gary, 1947—Diaries. 6. Natural history—Oklahoma. 7. Natural history—Great Plains. 8. Natural history—Cimarron River. 9. Oklahoma—Pictorial works. 10. Great Plains—Pictorial works. I. House, Don, 1951– II. Selman, Sue, 1948– III. Title.
F701 .L36 2002
976.6'053—dc21

2002012284

Frontispiece: Lena Ernest Selman and her family on the Ernest homestead near Freedom, Oklahoma.

To Brenda. Proof that behind every freelance writer there's a working spouse.

—Gary Lantz

I dedicate this book to my family, past, present, and future. To my grandparents, who journeyed in their youth to a land that was unsettled, uncivilized, and downright harsh at times. They came to love and respect this land and they passed that love on to their children and grandchildren. And to my parents, who taught me a deep respect and love for the land, wildlife, and the history of this great state. To my children, who have come to see the need to preserve this gift that has been bequeathed us. And to future generations, I can only hope that this book will serve as a tool and a reminder to preserve and appreciate their heritage. I would also like to thank Gary Lantz and Don House. Without their artistry and dedication this book would have been only a dream instead of reality.

—Sue Selman

To my father, Franklin, who taught me to keep my eyes open and a camera handy.

—Don House

Contents

Preface

Listening to the Wind: The Selman Ranch and Its Legacy

The Selman Ranch can, at times, seem as remote as the plains of Afghanistan. Yet with time and the familiarity that comes with close observation, this vast acreage shrinks to fit the ability of the mind to comprehend it, and one begins to understand the intimacy that a rancher shares with his or her land.

Intimacy with any landscape discloses nuances easily missed from a speeding car on some blacktop road. Through a windshield, the plains appear to be exactly what the name implies, a rugged landscape, tough as the leather in the men's favored footwear. However, when you get down on your knees amid the bunchgrass tussocks or ride a horse along the banks of Buffalo Creek, intimate observation reveals a complex system of plants that support a surprising variety of wildlife.

At first photographer Don House and I were drawn to the wildlife, as well as to the unique landscape carved by the Cimarron River and its tributaries. But after a while through our relationship with the Selman family we discovered a family history that rivaled the best of any television series or movie plot.

It became clear that we couldn't write about this ranch without including the people whose bones are buried here. J.O. Selman's spirit is as much a part of this landscape as the April flowers that signal the arrival of spring greenup, or the winter blasts that pile up drifting snow along the banks of Buffalo Creek. Pause for a moment amid the silence and wild beauty of the little stream called Sleeping Bear and it seems he's still there atop one of the stark gypsum buttes, sitting astride his little bay quarterhorse, watching everything that moves.

Watching with him, I believe, are the ghosts of the Cheyennes who once hunted and camped here. Spend some time on this ranch and it becomes apparent that the buffalo-hunting Indians and the longhorn-driving cowboys had much in common. They both valued individual freedom above all else; they understood that the health of the grass was the key to their simple economy; and they instinctively recognized that grasslands were meant to be big, that grazers were designed to roam. Big ranches survive because they are able to replicate, with their grazing regimen, the ancient patterns of the buffalo. J.O. Selman understood grass and disdained the plow. He was, like the Cheyennes before him, a true plainsman.

So this book grew into a tribute of sorts, hopefully a fitting and long overdue epitaph for the men and women who came here filled with hope, who fought stubbornly to stick it out, and who still live on in spirit . . . and in the stories that shall forever bear repeating at the mention of their names.

Our text is somewhat unusual in that it combines three distinct voices. I have done my best to capture a small part of the history of the region as well as offer insight into the nature of the Selman Ranch through the seasons.

In these same pages Sue Selman shares her memories of growing up on a traditional cattle ranch. Sue's plainspoken prose recalls when steers came north in rail cars, ranch wives cooked massive roundup meals on wood cookstoves, droughts brought a plague of grasshoppers to eat the paint off the barns, and kids dreamed of being cowboys like the whippet thin, sun-baked men who lived in the bunkhouse and who knew little else other than horses and cows, and maybe whiskey and women on a rare weekend when time allowed a trip to town.

The book wouldn't be complete without the visual experience provided by Don House, one of the best black-and-white photographers working in the Southwest. The photos speak for themselves, but in this instance readers are afforded a rare treat. Don agreed to unveil his superior writing talents as well, both in the meditations that accompany each picture and in the introduction and conclusion that establish a remarkable poetic, insightful tone for a book that in the end demanded such a voice.

So there you have it. A book of diverse talents and voices, all of them focused on the prairies that border Buffalo Creek. Please enjoy.

. . . Gary Lantz

Buffalo Creek Chronicles

Introduction

It was the cottonwoods that drew me out here. I was sitting four hundred miles away in a small cabin surrounded by trees when I received a call from Gary Lantz. He was researching an article for *American Forests* magazine that dealt with the trouble prairie cottonwoods were having trying to survive in a region whose aquifer was dropping and where invasive alien species were competing with natives for what little was left. He needed a photograph. I remember saying this:

"But, Gary, it's the middle of winter, there are no leaves."

So on a freezing cold December morning, I drove out of the Ozark valley I call home, through red oak, white oak, chinquapin oak, hackberry, sycamore, cedar, maple, ash, serviceberry, elm, sweetgum, redbud, sassafras, beech, paw-paw, hickory, haw, and walnut, and headed west toward the Oklahoma prairie that a neighbor of mine, who had once lived there, described this way:

"If there's a tree and it ain't growing on a river, it ain't there."

I stopped in Norman, just south of Oklahoma City, to pick up Gary, then continued northwest to Woodward. As we passed a cup of coffee back and forth over a mound of camping gear and camera bags, he filled me in on our destination—Sue Selman's ranch. I kept interrupting him:

"Fifteen *thousand* acres?"

"But, Gary, that's over twenty square miles."

"The Cimarron River runs right through it?"

We drove in relative silence after that. I was trying to put the numbers, the scale, into perspective. In the rugged Ozarks, an 80-acre farm is the norm, 160 acres considered big, 320 huge, and a full section of 640 almost unbelievable. I was also thinking of Gary's last comment:

"It's small by prairie standards."

A few hours later I was wading through fragrant sage along the banks of Buffalo Creek, a tributary of the Cimarron, and listening to a litany of the Latin names of plants I didn't recognize. Gary Lantz is my idea of the perfect hiking companion. He will travel in silence when you want silence, lost in his own thoughts, stopping frequently to make notes in his journal, or to photograph a flower for later study, but if you give him permission by asking a question, he can identify every plant by scientific and common name, tell you if it's native or alien, and describe how it fits into the overall ecology of the prairie. That depth extends to mammals, insects, and birds as well. When we sit to rest pack-weary backs and I close my eyes to listen to the sounds around me, the noises he makes fit in naturally, a little scratching of pen on paper, a slight click of a shutter, a long sigh of delight at having found a rare plant, or my favorite—a low mumbled, just audible, "Who the hell are you?" when a flower or a flap of wings momentarily stumps him. The photograph that resulted from that trip is on page 105. Standing in the presence of those trees, the seeds of this book were planted, then nurtured with warm coffee and biscuits in Sue's kitchen a few hours later. It was clear to all of us that one day and one story would not do it justice.

Meeting Sue for the first time can be a humbling experience. She would be embarrassed to have me say that, but it fits. Rancher, daughter of a rancher, and granddaughter of a rancher, the energy and determination and physical strength and discipline that she brings to bear in running the business is the humbling part. It makes my daily routines and procrastination-plagued projects seem ludicrous. But spend some time around her and there is also an empowerment as you feel your own attitude changing, as if a message has been telegraphed to your subconscious:

"You can do anything, so get off your butt and do it."

We had talked for an hour, that first meeting around her kitchen table, before it struck me how unusual the conversation seemed. I was talking to a rancher, yet she had not once bemoaned government regulations, lamented stupid tree huggers, or spit on the floor at the mention of endangered species. Her driving consideration seemed to be how to make a living on this land without destroying it in the process. And wrapped up in that determination was a clear sense of her place in the continuing history of the ranch, from her parents to her grandparents to the native Americans whose presence on this ranch stretches back further than her box of yellowing family photographs.

The prairie is *powerful* and *intense.* That is the impression that I take with me. Its human history, natural history, climate, and topography. It is absolutely *beautiful.* And like nowhere else I know, it can make you feel absolutely *alone.* Through the eyes of Gary, naturalist and wordsmith, and Sue, rancher and businesswoman, I have been able to see the prairie with greater understanding. It is so much more than just learning the names and niches of flora and fauna, how to distinguish between overgrazed and healthy pasture, or what combination of plants will guarantee a diverse wildlife population. It has been an opportunity to open a door and see into the past, and then to open a second door and see what the future might hold if people who love the prairie are allowed to do what needs to be done. I have been given a gift, and with Gary's insights and Sue's remembrances, we pass it on to you.

. . . Don House

Jimmy Few Clothes

On the Trail to Greener Pastures

by Gary Lantz

Photos of a prosperous J.O. Selman in his ranching heyday show a man unusually well dressed for a working cowboy. His grandchildren remember that he was almost fastidious in his attire, even on horseback.

However, fellow cowhand J. N. Milhollon recalled another James O. Selman. The J.O. he remembered was a ragged, scruffy kid desperate for work, willing to do anything to escape the poverty of a west Texas farm.

Milhollon was trail and roundup boss for A. H. Tandy's Bar Z Ranch, headquartered near Haskell, Texas. In the spring of 1892, young Selman rode into Haskell, hitched his pony to the courthouse fence, and walked across the street to where Milhollon and other cowboys were gathered on the sidewalk.

"He was just a boy, fifteen years old," the Tandy trail boss recalled. "He was riding a little sore-backed bay with a blaze face, and he wanted to know where he could find a job."

Milhollon and his crew were gathering a herd of longhorns to deliver to Henrietta, Texas, and he invited Selman to come along as the outfit's horse wrangler. "He was mighty glad for that offer," the old drover recalled during the 1948 interview.

"Jim was a ragged little cuss, but he was game," Milhollon declared. The boy told his new boss that he'd left the home farm in Throckmorton County because he couldn't get along with his stepmother. After a few days on the trail the older cowboys grew fond of young Selman and searched through their belongings for spare clothes for the nearly threadbare new hand. Milhollon remembers that the drovers outfitted their rookie wrangler with overalls, shirt, socks, vest, and "other necessities until he was pretty well rigged out."

One of the cowboys observed that "poor little Jimmy few clothes, when I first saw him he had scarcely any clothes on his back. But now his back is mostly covered with clothes." From then on J.O. Selman's nickname would be Jimmy Few Clothes, and he kept it with pride until his death.

Heir to the kingdom. J.O. Selman's son Bob learns about ranch life at an early age, riding behind one of the Selman Ranch cowboys.

Trail boss Milhollon recalled that J.O. worked for the Bar Z outfit for a number of years and "made a good cowhand, a man you could depend on either on the range, the trail, day herd or night herd, good weather or bad. He stayed with the cattle until somebody relieved him. That was one of the things that impressed us boys: in those days you had to stay with the cattle above all else."

"On one trail drive Jimmy and I were out on the prairie together. It was in 1894 and we were moving a herd from down in Texas to north of Woodward, Oklahoma, where Mr. Tandy had established a ranch," Milhollon explained. "Late one evening we threw the herd in on Commission Creek south of Higgins, so they could water. We bedded down in the valley north of the creek, and for some reason the cattle were restless. We had to double the guard to prevent a stampede."

The next day the crew turned the herd back out on the trail, pointed the wagon toward Higgins for a supply of groceries from W. F. Peugh's store in the little Texas Panhandle frontier town and managed to push the longhorns as far as Rock Creek, near modern-day Shattuck, Oklahoma.

"There wasn't a house between Higgins and where Shattuck now stands," Milhollon remembered. At the time Shattuck was nothing more than a railroad switch and a small store operating out of a boxcar.

"When we finally arrived at Woodward we found a shack town with only one street a mile or more long," Milhollon added. "There were nine saloons all running on high. Finally we headed the cattle northeast out of Woodward and turned them out on Sand Creek. As soon as we were done, Jimmy headed back down south to come up with another herd."

"I introduced Jimmy to some girls, but I had to hesitate because I couldn't remember his real name," Milhollon reminisced. "But the boy spoke up and said, 'My name is Few Clothes.' I let it go at that; I think he was really proud of his nickname."

Another "young Jimmy Few Clothes" story has it that bossman A. H. Tandy determined that Selman was too young for the trail and fired him after he returned from wrangling horses on the cattle drive to Henrietta. However, Tandy was barely out of sight before range boss Milhollon rehired little Jimmy.

"This happened several times," one old cowhand recalled. "Finally Old Man Tandy relented and not only kept the kid around but also began to take him under his wing."

Tandy was a tolerant man, evidenced in an exchange between the cattle baron and the destitute kid, who, according to one of many legends, told the rancher he'd be content to work for a few clothes, a plug of tobacco, and the use of a good saddle horse if he'd just keep him on as a cowhand.

Early one morning Tandy ordered Jimmy out to ride fence. He was back at noon in time for dinner.

"Did you ride the fence like I told you?" Tandy inquired.

Selman replied that he had.

Then the rancher asked his smooth-cheeked cowboy if the fence was in good condition.

"No sir, it was in damned sorry shape," young Selman answered.

By the time he married Lena, "Jimmy Few Clothes" Selman was rather well dressed, as evidenced by this formal portrait of the young couple.

"Well, did you fix it?" Tandy demanded.

"No sir. You told me to ride it. You didn't tell me to fix it," Jimmy muttered as he reached for a coffee cup.

That's when Tandy turned to the rest of the cattle crew, speaking loudly enough so that all, including Selman, could hear. "Don't anyone tell Jimmy when dinner is ready. Let's see if he knows enough to eat without being told."

Obviously Jimmy Few Clothes was a fast learner, for two years after he signed on with the Bar Z crew the teenager was moving vast numbers of longhorns north to Tandy's ranch between the North Canadian and Cimarron Rivers in the old Cherokee Strip country in what is now northwest Oklahoma. In the spring of 1894 Selman and his trail crew delivered 2,000 longhorns to the Cimarron range, then turned the chuckwagon and horses back south to gather another herd. By August the drovers returned with 3,400 more Texas cattle to graze on a 20-square-mile pasture bordering the banks of Sand Creek. Selman soon fell in love with the gypsum breaks south of the river, a wide and rolling land carpeted with good grass and drained by several spring-fed streams. Jimmy told his friends back in Texas that the Sand Creek spread was the finest cattle country he had ever seen.

In time Tandy would move his headquarters and home to Woodward, where Selman, now a veteran hand within the Bar Z operation, would help his boss and mentor build one of the city's more fashionable houses. Just as Tandy's fortunes grew along the banks of the North Canadian, the young Throckmorton County cowboy's opportunities soared as well. Selman began to acquire rangeland of his own—a place near the confluence of Beaver and Wolf Creeks, a strip of land along Buffalo Creek that included portions of the Sand and Sleeping Bear Creek drainages, pasture along the rugged Trader's Creek breaks, some good grass near a cowboy community perched on the north bank of the Cimarron, a little town called Freedom.

Back home in the rugged Brazos River country of west Texas, the hardscrabble Selman family wasn't faring as well as the formerly desperate son who had fled the crowded cabin with only a few ragged clothes on his back. Letters to J.O. from his father describe with evangelical flare the torment following years of drought, of shriveled grass, of wheat and corn withering in the fields.

The threat of starvation compelled J.O.'s brothers to scatter over the plains, seeking cowboy work in the Texas Panhandle, New Mexico, even up into Montana as stay-at-home family members scratched for food along the riverbank in the wretched residue of what they'd hoped would be their pastures of plenty. The elder Selman's letters, richly spiced with scripture, also reflect his envy of the good grass and water accumulated by his ambitious son. Several letters express a desire to make another, hopefully final move up to the country along the Cimarron or maybe to some good grazing country in the

Texas Panhandle. Even so, the drought-stricken, grass-poor, flooded-out farmer laments that "what available land that is left is filling up fast." And so it was.

One poignant piece of mail announces the return of rain to the parched countryside, rain pounding the earth with such force that the river rose quickly and threatened to carry away the family's ramshackle home. In desperation the remaining Texas Selmans tore their house apart plank by plank and carried the boards uphill. However, all they could do was watch as the bank, denuded by years of overgrazing, timber cutting, and plowing every spare inch of the bottomland, quickly eroded away from under the stone structure and a muddy torrent carried their chimney downstream. As they worked they more than likely remembered another flood that swept downriver just as J.O.'s father was attempting to move his first wife's coffin across the river to a neighborhood cemetery. The rapidly rising waters floated the casket out of the buckboard and down the channel like a rudderless boat. The funeral couldn't proceed until several of the men caught up with the crude wooden box and swam it across to the gravesite.

The significance of the Texas tragedies back home weren't lost on the young cowboy who'd vowed to make a better life for himself in the sandhills and sage along the Cimarron. A small man who rode a tough little quarterhorse, J.O. always stood just a little in his stirrups, possibly to elevate the crown of his hat an inch or two. However, J.O.'s potential financial stature needed very little boosting in the eyes of area banks, and before long Selman was entering into profitable partnerships and amassing a spread that would eventually total some 60,000 acres in the Buffalo, Sand, and Sleeping Bear Creek prairies he'd grown to love.

J.O.'s growing reputation as a prosperous rancher also increased his ranking as one of the area's most eligible bachelors. Therefore the cowboy, by then noted for his fondness for and fine taste in clothes, found time amid his empire building to court the local ladies. One in particular caught his eye, a smart and lovely woman in her early twenties named Lena Earnest, a farmer's daughter who taught school in the rugged ranch country a few miles east of J.O.'s Buffalo Creek pastures.

An early picture of Lena reflects the rawhide grit required of women who managed to both survive and prevail in this parched and windswept landscape. Her face is chiseled, handsome, dignified, with intelligent yet piercing eyes. However, the young

Lena Selman taught at this prairie schoolhouse.

woman's attire would warrant an admiring whistle from a muleskinner—a fur coat that appears to be straight off the shoulders of a shaggy buffalo, plus a rifle tucked in the crook of her arm in a manner that makes it seem quite natural there. This, the picture declares, is a woman to be reckoned with, a woman with enough steel in her soul to survive drought and blizzard and still keep a working ranch in the ledger book's black.

So J.O. came calling at the Earnest household, although his courtship with Lena was colored by a tint of scandal. Lena was one of those rarities that other women on the frontier loved to whisper about: a divorcee, and a bizarre one at that. Her family had moved to the Cimarron country from St. Louis when Lena was around sixteen. Bad

Lena Selman was as tough-minded as she was pretty. The carbine in this studio portrait, matched with a formal length fur-trimmed coat, clearly gets the point across.

investments stripped Lena's father of both his money and his reputation, so the family filed a claim on some land in the Indian Territory and vowed to start over again. The covered wagon ride nearly finished the fragile family, but even this harsh form of frontier torment couldn't equal the despair they felt when the horses halted at the edge of their eagerly anticipated new property line. Due to an administrative error, another pioneer family had filed on the same property and were already living there in a dugout, the typical plains home of the time that amounted to little more than a ditch carved deep into a hillside and covered with a roof of planks and sod.

A bitter plains winter was already bearing down on the gypsum canyons bordering the Cimarron, so Lena and her family had few options other than to accept an offer to share the dugout until spring, when the disputed boundaries could be clarified. Considering what was at stake, it is difficult to imagine such an act of hospitality today. Yet the two families managed to cohabit under these harsh and obviously emotionally trying conditions, and the claim boundaries were resolved to the satisfaction of all before planting time arrived.

Lena Earnest Selman's family left St. Louis believing they held title to a homestead in northwest Oklahoma. Upon arrival they discovered the land was already occupied, and shared a dugout with the interlopers until the legalities could be worked out in the spring.

It must have been the epitome of culture shock, coming from St. Louis opulence to a home dug into the side of a hill. However, the Earnest family made the adjustment and even bright, pretty Lena seemed content with her lot—until, that is, a young traveling salesman, or "drummer" as they were known, knocked on the door of the frame home they'd built, following months of living beneath the earth.

Many years later Lena would tell her granddaughter that the salesman was the most handsome man she had ever seen, and his quick smile and glib ways were enough to persuade her, when the invitation came, to gather up her belongings, board a train with the stranger, and never look back. They were married at the first major city and traveled on to California, while back in Oklahoma Lena's parents anguished over their daughter's unknown fate.

The giddy young woman was in store for some severe emotional trauma soon after she unpacked in her new husband's home. Suddenly the charm vanished and in its place

Lena Earnest and family before Lena became Mrs. J.O. Selman. Lena taught school while her family proved to be tenacious farmers in a land more suitable for cattle grazing.

Lena discovered a domineering man who locked her in the house each morning as he left for work. Spirits sank as her husband's violent paranoia grew, and finally Lena managed to sneak a letter back home. Please help me, I'm a prisoner here, it said. She knew her mother would be sympathetic, but her father was a stern man who found it hard to forgive. This time, however, he read the letter, reached in his pocket and pulled out enough cash to purchase his daughter a train ticket home. When the time was right, Lena packed her bags and fled.

In the years that followed the schoolteacher was courted by a host of suitors, and she confessed to her mother that she was torn between her feelings for two men—one of them the rancher James O. Selman. The women were washing dishes when Lena poured her heart out, and her mother just smiled and said, "Well, you'd better start making up your mind, because I believe providence has answered and that's Jim Selman standing at our door."

At that moment J.O. had some serious matters on his mind as well. He'd hitched his team that morning, intending to drive the horses to the Earnest household and convince Lena to marry him. A cowboy who met Selman on the road asked the rancher where he was headed. J.O. grinned and replied, "I'm gone in search of a lost heifer." A few months later the bright and lovely Lena and the rags-to-riches Texas cowboy were wed.

J.O. and Lena built a house on Buffalo Creek, in a sheltered valley amid tall cottonwoods and later, under Siberian elms planted to provide shade during the stifling summers. It was a large frame house, with ample porch space covered with screening to keep out the swarming flies. Behind the barns and outbuildings were the AT&S railroad tracks that paralleled corrals and shipping pens. By now the cattle drives of J.O.'s youth were only memories, replaced by massive engines and railroad cars hauling steers north from Texas to fatten on Selman grass, then for the final ride to the meat packers.

Today the weathered house and outbuildings serve mostly as a refuge for birds, wood rats, and the occasional cow seeking shelter from a norther blowing icy sleet over the sandhills bordering Buffalo Creek. Even so, with imagination you can still sense the beauty and tranquility of the place, this sturdy, lively headquarters for a prosperous and expanding cattle business. The transplanted elms are huge and quite tall now, with branches succumbing to age and disease, and the railroad tracks are gone, even though the shipping pens appear sturdy, still capable of corralling beef bound for Kansas City stockyards.

Lena Selman proved to be the cornerstone of J.O.'s cattle operation. She was a good hand outdoors as well as an excellent bookkeeper, and little Bob was content to ride along no matter the chore at hand.

But that's a scene that only ghost riders revel in, because the silence of this place will never again know the piercing whistle of a train. Today's cattle depart these pastures in trucks bound for feedlots where thousands are squeezed together to fatten on ample rations of grain. Only a few old cowhands remember the din of agitated steers, the mechanical shriek and lurch of the train, the shouts of the cowboys as they sorted the new arrivals spilling out of the ventilated cattle cars. The bawling steers were almost deafening as they churned up mud following an overnight thunderstorm or raised a cloud of dust if April remained dry.

Lena and J.O. moved to Woodward when their children, Ilene and Bob, were old enough for school. J.O. continued to manage some 60,000 acres of cattleland along with a growing list of business endeavors, all under the watchful eye of his intelligent and tough-minded wife. Later he'd invite his brother Green Selman to take over the ranch on Trader's Creek. Granddaughter Billie remembers the warmth that emanated from her grandparents' home, and the Indian pony that J.O. gave her, a gift he purchased from some local Cheyennes. "If you're going to ride an Indian pony, then you're going to learn to ride like an Indian," J.O. commanded. So the girl rode bareback, trailing her beloved grandfather as he surveyed his acreage at a steady trot. Occasionally they'd take a break by a prairie dog town, where the aging cowman would rest for a minute in the saddle and chuckle at the antics of the increasingly rare critters that most cattlemen have come to despise.

Today prairie dogs are gone from the Selman Ranch, and only a scattering of mesquite trees remain as a reminder of the infestations that followed heavy grazing when ranches here were young and their ability to provide nourishing grass seemed limitless. Years of too many cattle and too little rain allowed the invasive mesquite to gain a foothold and spread across portions of the ranch. Finally the Selman heirs decided to eradicate as many of the trees as they could.

It was a balmy spring, perfect for planting, and young Billie thought that a wonderful surprise for her Uncle Bob would be a shade tree, planted in the yard. The girl convinced her aunt to help her dig up a small yet lovely tree with lacy foliage she particularly favored. Billie brought the specimen back from the ranch and proudly planted it near the front door.

She waited by the porch, ready to witness her uncle's pleasure. But when the tired and dusty ranch boss saw the spindly bush, neatly pruned and freshly watered, he almost choked on his words.

"I just spent a fortune killing about a million of those things," the rancher snorted. "And now you've gone and planted one in my front yard!"

Mesquite trees remain a part of the modern Selman Ranch flora, but nowhere are they invasive or troublesome, a testimonial to improved range management and modern grazing practices. In fact some of the older trees, gnarled and twisted, lonesome and aloof, add character to the broken ridges overlooking the Cimarron. These Methuselah trees supply a splash of soft green foliage to accent early spring flowers, or serve as a perch for a redtail hawk or as a scent post for a coyote with hungry pups to feed.

The J.O. Selman ranching era commenced with a cattle drive and concluded in much the same manner. Around 1947, J.O. decided to swap some 800 head of registered cows for about 1,000 two-year-old steers owned by the Barby Ranch, a big spread to the northwest in the Oklahoma Panhandle.

Selman Ranch cowboys working cattle the old-fashioned way, when skill with a rope and horsemanship still mattered.

The elder Selman and a crew on horseback started his cows up the North Canadian River. Higher up on Beaver Creek, the Barbys started their steers south.

The nights were cold, and a couple of younger Selman kin, along for their first trail-driving adventure, awoke to find frost on their blankets. They discovered that J.O. had also felt the sting of the night air and crawled beneath the blankets the boys shared—in the middle, of course. The youngsters clung to the edge of their bedroll and absorbed a teeth-chattering lesson in how old-time cowhands lived on the verge of heatstroke or frostbite, with only a Stetson or a wool blanket to intervene.

The drive took seven days, and the men were fed from a pickup truck outfitted as a chuckwagon. The two outfits met near the small northwest Oklahoma town of Gate, at the foot of a big sandhill that served as an area landmark.

Branding calves under J.O.'s watch required the ability to throw a tight loop as well as rawhide toughness.

The sandhill also served as a dividing line. Barby's steers were herded to the east side of the prominence, Selman's cows to the west. Soon the branding irons were hot and the range reverberated with the curses of cowboys and bawling cattle. It was a scene straight out of J.O.'s trail drover past, and he must have watched the activity with both sadness and satisfaction. Soon he'd leave all the ranching duties to his son and settle into the less strenuous life of managing varied business interests in town. For "Jimmy Few Clothes" Selman, this final cattle drive had been his way of concluding a colorful era, an encore for a long-running western drama written in mud and blood, hailstorms and drought. When J.O. turned his newly acquired steers toward home, he knew the curtains were closing on a way of life he'd cherished. Standing just a little in the stirrups, the tough old Texas trail hand eased his horse into a trot. He'd literally lived his life from rags to riches. And no matter what the task, he could say with satisfaction that he'd always played his part.

≈≈≈≈≈≈≈≈≈≈

The written history of the prairie country bordering Buffalo Creek stretches all the way back to 1599, when Spanish explorer Onate camped beside one of the streams flowing into the Cimarron River and noted in his journal that the surrounding plains were covered with innumerable wild cattle, or buffalo.

Two centuries would pass before another white explorer ventured up the Cimarron to the mouth of Buffalo Creek. In June of 1811, American adventurer George Sibley came in search of the Rock Saline, a large deposit of salt at the time known only to wandering bands of Indians.

Sibley was intrigued by tales of this great mountain of salt and decided to see for himself. He commissioned a party of 90 Osage warriors to join him on a trek through the rough gypsum breaks of what is now southwest Kansas and northwest Oklahoma. The easterner was captivated by the rugged beauty of the countryside, and he wrote that the red clay buttes capped with white rock reminded him of the "magnificent ruins of some ancient city."

Sibley also noted that while passing through one of the deeper canyons, both he and his Osage escort were suddenly silenced by the strange reverence of the place. They continued on, but without speaking a word. A short time later the trail entered sagebrush-covered sandhills, where the explorer claimed he discovered a herd of some 30,000 buffalo watering along a small stream. The Osages gave chase, and Sibley said the narrow stream valley reverberated like thunder as the buffalo fled. The crack of gunfire could be heard over the pounding hooves, and afterward Sibley counted 27 dead buffalo and two injured Osages.

The trail proceeded to the summit of a bluff offering an expansive view of the glistening salt plain of the Cimarron. Sibley reported that this natural wonder, viewed from a rocky outcrop wedged between the river and the mouth of Buffalo Creek, revealed itself as a shimmering expanse of red sand and salt. The adventurer estimated the size of the salt deposit to be somewhere around 500 acres, but in fact it was more like 5,000.

The prairie traveler's notations reflected his admiration for the high bluffs bordering the river, bluffs he said swarmed with nesting swallows. Sibley also mentioned salt springs that flowed from beneath the bluffs and spilled out onto the vast flats bordering the Cimarron. The Osages told him that during the summer drought the salt crust grew to be five to eight inches thick, and that large volcano-shaped cones formed at the mouths of the salt springs.

Almost thirty years later, in 1839 and 1840, a wealthy young Frenchman, Victor Tixier, intrigued with American Indians and seeking adventure, would join another Osage hunting party bound for the buffalo plains bordering the Cimarron. He, too, would praise the beauty of the land bordering the valley of the Cimarron and marvel over the abundant wildlife.

The buckskin-clad nobleman was well educated, and his romantic notions pertaining to all things Indian and the American West didn't taint his carefully inscribed observations.

"We arrived at our last camp toward July 20," Tixier wrote of country near the Salt Plain of the Cimarron. "The prairie was arid and rolling, the horizon bordered by high ridges. Several rivers shaded by beautiful trees were winding between the hills. At every step one encountered dried salt marshes on which the crystallized salt had left a white crust, which our horses licked while walking. We camped in a strong location well protected by deep ravines. There was a great abundance of game on the plains: antelope and Canadian stag (the latter in this country is confused with the elk, which has become very scarce), buffalo and deer were seen frequently, and black bears and grizzly bears dwelt in the deep valleys covered with thickets."

Three more years would pass before Captain Nathan Boone, son of legendary frontiersman Daniel Boone, would lead a troop of 60 dragoons up the Cimarron on a reconnaissance mission. He crossed the river near the Selman Ranch on July 1, 1843.

"Broke up camp and moved across the river and salt plain SW four miles and encamped on a small creek of water slightly brackish, but as good as any we have found in the gypsum," Boone noted in his journal. "In crossing the river we had some trouble in quicksand with the wagons. The bed of the river and salt plain is about two miles wide. This afternoon it rained and Talle, the Osage chief, and his people came and encamped with us. They informed us that all the crystallized salt on the surface of the plain is washed off when it rains heavily and that in a few days the sun will bring it back."

Boone and his men pitched their tents on Trader's Creek and spent the second day of July exploring the area. "Parties went off to hunt buffalo, and numbers were killed near our camp," Boone wrote. "A party went to explore the salt plain and search for rock salt. The whole cave on the right of the two forks of the river appeared to be one immense salt spring of water so much concentrated that as soon as it reaches the point of breaking forth, it begins depositing its salt."

Seventeen years after Boone explored the Cimarron saline, U.S. relations with the indigenous Comanche, Kiowa, Plains Apache, Cheyenne, and Arapaho had disintegrated into open warfare, and cavalry troops patrolled the rugged country where the Cimarron and Buffalo Creek merge. One patrol reported spotting four bears, several deer, and plenty of buffalo in the area and commented that the grass was better than any they'd found since leaving the rich valley of the Arkansas River. And in 1869, Civil War hero General Phillip Sheridan, in the area to track down and punish Cheyenne raiders, reported a turkey roost stretching for nearly three miles along the cottonwood-studded banks of the North Canadian River.

The abundance of wildlife along the prairie waterways was certainly something to marvel over, as was the speed of the creatures' impending demise. In 1877 hungry members of the Cheyenne and Arapaho tribes journeyed to Fort Supply, a military outpost near the junction of Wolf Creek and the Beaver River. Later much of the area would become Selman grazing land, but at the time it was still Indian territory.

The Cheyennes hastened to the country between the Cimarron and North Canadian due to reports that a herd of buffalo could still be found in the area. However, they weren't prepared to meet their bitter enemies, the Pawnees, who had left their reservation east of Cheyenne country and ridden west to the military camp in a desperate search for food.

The Pawnee chief reported that his people were starving. He'd decided to risk a confrontation with the tribe's traditional enemies rather than watch children go hungry. Without meat, they were certain to die anyway.

The commander at Fort Supply kept the tribes separated and brokered a deal. Each tribe was assigned a direction to hunt, and the hunters took an oath not to deviate from this course and risk the chance of armed confrontation.

The diplomacy worked. Unfortunately the hunt didn't. Few buffalo were to be found, and the Indians wandered back to their respective agencies subsisting on survival rations issued by the post commandant. What they didn't know was that only a few months before this final autumn hunt, an estimated 40,000 buffalo were grazing along the banks of the North Canadian. Then the white hide hunters came, along with Indians from other tribes. The killing was little more than an intense business transaction, without a thought given to conservation or the plight of local tribes who needed the animals for meat. The slaughter would not cease until the last buffalo lay dead.

The Pawnees, chewing on their Army-issue hardtack biscuits as they retreated to their agency on Black Bear Creek, must have wondered how their once-splendid world could evaporate so rapidly. Before the white man came the Cimarron salt springs had been a place for spiritual renewal, and Pawnees from Nebraska, Caddoes from the Red River country to the southeast, and even Arikaras from the Dakotas congregated here to hunt buffalo, gather salt, and make things right with their God. According to Pawnee legend an ancient red cedar grew from the base of the bluff where Buffalo Creek and the Cimarron converge. This was a twisted, gnarled old patriarch tree, an altar to hear the priests' prayers and a scaffold to receive sacrifices. But the rancid bacon from the fort was hard to swallow in the place of sweet, nourishing buffalo, and there was little reason to stop and pray, little to envy in a sedentary life lacking the hunt. So they left their proud old gods there on the prairie and rode back to accept a diseased, culturally fragmented, whiskey-soaked fate.

Below a meandering beaver pond, near where a huge white slab of gypsum has weathered away from a tall bluff, the currents of Sleeping Bear Creek have uncovered buffalo bones buried in a bed of pale sand.

Enormous, yellow with age, these bones bear silent witness to the days when Cheyennes camped here, along a stream that caught the waters of countless springs, a place of shade and solitude and good grazing for herds of horses, near where the buffalo came for salt.

Historical records indicate that shortly after the Civil War most of the southern Cheyenne people were encamped together along the banks of Sleeping Bear Creek when a U.S. Army scout, himself a mixed-blood Cheyenne, crossed the Cimarron and rode amid the tipis, there to do the war department's bidding.

The scout said the "great white father" wanted to negotiate a final treaty on the banks of Medicine Lodge Creek, and that the people should come to this place and listen. So the Cheyennes packed their belongings and rode north to where Elm Creek empties into Medicine Lodge Creek. Overhead, the migrating flocks of late October, shorebirds, geese, and sandhill cranes, voiced a wild and haunting dirge for the procession.

Today we know that the Medicine Lodge Treaty of 1867 was nothing more than a bribe—a guise to rid Kansas of plains Indians hostile to the tidal wave of settlers attracted to the fertile prairies. Some 15,000 Cheyenne, Arapaho, Comanche, Kiowa, and Plains Apache were told that if they stayed south, deep in the Indian Territory, the lands they accepted there would be kept free of sodbusters and buffalo hunters.

In truth, some of the last free days of the southern Cheyenne were those spent in the Sleeping Bear Creek campsite. One can only hope that the buffalo bones that currently protrude from the banks are from an animal that provided repast for a feast, and that the old men smoked and dreamed of a lifetime of hunts across prairies reaching from the upper Canadian River to the salty Red, and not of the impending rattle of sabers that would slash men, women, and children indiscriminately and leave a record of America's shame in the prairie snow.

≈≈≈≈≈≈≈≈

Lena and J.O. Selman had two children, Ilene and Bob. Ilene was a restless child, and as soon as the girl was old enough she drifted away with the prairie winds. Bob, on the other hand, was much more solidly set in place. When the elder Selman decided it was time to retire from active ranching, he told his son the land was his—Bob could have as much as he felt he could pay for. The younger Selman purchased the heart of his father's holdings, the prime acres including the core of the Buffalo, Sleeping Bear, and Sand Creek drainages.

Bob Selman was a burly youth, a man born into the cattle business which he served dutifully, even though his real love was the heavy machinery parked beside the barn. While other ranchers watched their bank accounts dry up when drought withered the grass, Bob found work and made good money preparing drilling sites for the numerous oil companies operating in the area, or building ponds to capture the precious groundwater that seeped out of the sandhills.

Bob lost a finger when he slipped while coupling a stock trailer to a truck. Rather than regard the accident as a catastrophe, the young rancher picked up the severed appendage, placed it in a bottle of alcohol, and kept it as a trophy.

This was the way men reacted to pain and loss during the hard years of the Depression, and the Selmans took pride in their toughness. Some of it was certainly genetic, but much of the attitude was cultural and environmental as well. As lovely as this land could be after a May thunderstorm, a sea of green laced with spring flowers, all who lived here knew it was rarely an easy place for more than a day. Bob Selman was successful because he knew how to be hard when hard times called for it. He'd learned it from J.O., who gave him little quarter in that regard. But maybe more than anything else, Bob Selman raised a family and made a living off of fickle native grassland because some powerful ghosts were always looking over his shoulder, ceaselessly demanding that he get it done.

Rangeland cattle ranching never has been a business you choose from a college curriculum. Instead it's a matter of blood. Ranches are built over generations, and many are lost in a matter of weeks or months or years when a weak link appears in the chain of heirs. Shakespeare would appreciate the feudal aspects of cattle ranching; the power of a name, the image of a brand that symbolizes human genetics maybe more than that of the bovine stock, the way that the slightest character flaw, individual or family, is scrutinized and over the years immortalized as a matter of regional cultural lore.

Bob Selman's biggest weakness also proved to be his greatest strength. Her name was Wilma, and she was the daughter of a cowboy-turned-farmer who settled on land along the Cimarron River maybe five miles from the Selman ranch house on Buffalo Creek.

The original prairie tractor.

Wilma's father, John, was the son of Wiley Cowan, a frontiersman who shared campfire coffee with legendary westerners like Kit Carson and herded cattle in the Cimarron country while buffalo hunters were still about their bloody work. The family remembers John Cowan as a good cowboy-turned-farmer who fought an unenviable battle to grow a living out of parched earth. Eventually the Cowans moved to the community of Selman not far from the original ranch headquarters, where Wilma's mother, Eula, taught school.

Wilma was a cowgirl both in blood and ambition, a rodeo queen who understood horses and also how to wrangle the top cowboy catch in the region—Bob Selman. The couple eloped in 1931, when Wilma was only 16. The news was not met with favor in what was by then the fashionable and politically connected town home of J.O. and Lena Selman.

However, this slim girl who could flash radiant good looks proved to be both headstrong and tough. Some feel that Wilma was the progeny most like her grandfather Wiley, a cowboy's cowboy and local legend. The girl worked hard, she had a natural

Wilma Selman was both a hardworking ranch wife and a star in her own right. An accomplished rider, she was the sweetheart of many a cow country rodeo.

instinct for handling horses and cattle, and she seemed fearless. Wilma may have been the best cowboy on the Selman Ranch, but in the 1940s women were cooks and housekeepers first. Even then, however, wives and daughters oftentimes assumed men's roles when necessary, simply because there was much work to be done and someone had to do it. But a rancher's wife had to first feed cowboys before she could labor beside them. Wilma was capable of both.

When Wilma's rodeo queen portrait appeared in some eastern newspapers, the young woman received written proposals of marriage from as far away as New York. But all she wanted was Bob Selman . . . and maybe the land, cattle, and heritage that came with the name. The marriage was stormy yet managed to last: once, after a particularly bitter fight, Wilma left for California, leaving Bob in a besotted emotional state that proved unshakeable until J.O., realizing that his son's depression was real and destined to endure until he and his bride were reunited, journeyed to the West Coast and brought the girl home.

Bob and Wilma Selman worked hard, played hard, and kept the ranch together through the Depression and the successive dry years that withered the southern plains. They watched the dust clouds rip the topsoil from a region that should have remained in native grass, protected by plant communities that evolved to withstand the periodic droughts and the blistering sun. But at the time settlers knew only to plant, then pray for rain.

Selman rangeland survived because it had adapted to survive, but still the curse of the thirties and another major drought in the fifties brought plagues that were biblical in proportion, plagues like swarms of grasshoppers that stripped trees of leaves and then ate the paint from the outbuildings. There were times when the Selmans were forced to load their cattle onto railroad cars and ship the animals north, literally to greener pastures.

But the ranch survived, once again in part due to the brand and all it symbolized. Lesser outfits starved out during the years of dust, drought, and Depression, and J.O.'s granddaughter Billie remembers how the homesteaders, the financially and emotionally ruined dirt farmers, would stop by the grand Selman home in Woodward, their downcast eyes relaying their mission before they could even find the strength to form the words.

Billie said J.O. would ask each of the men if a loan would help them hang onto their land, but by the time the weary farmers arrived on the Selman porch they were thoroughly defeated, and most would reply, "I just can't make it anymore." So J.O. would arrange to purchase the property. Some would accuse him of "stealing" land at dirt-cheap

Wilma Selman continued to make headlines for years following her marriage to Bob. She was a headliner at the annual Elks Rodeo and her picture circulated widely, soliciting proposals from as far away as New York.

Sue Selman learned quickly that a budding cowgirl didn't have much time for dolls.

prices, but Billie said her grandfather displayed genuine concern, paid what he could, and that the families who sold out were generally on the verge of starvation and appreciated whatever cash they could accumulate. Soon they'd be on the highway, headed toward cities where relatives lived and the possibility of available jobs; farmers became autoworkers or oil field roustabouts or wandered to migrant camps in California, awaiting their turn in the orchards, work that amounted to little more than white slavery.

Reputations take on a life of their own as time passes, and the reputation of J.O. Selman is not as sweet to some. Oldtimers in the area will tell you that Selman was little more than a robber baron and a tyrant, ruthless when it came to squeezing sodbusters off their land. In truth family members will be the first to affirm that the patriarch of the clan had little use for dirt farmers and placed his faith in what the land provided naturally, which was native grass. He knew that small 160-acre homesteads, dependent upon rain for a meager cash crop with precious little forage left over for a horse and a few cows, was labor-intensive suicide. In this land of extremes, one mild growing season with adequate rain would be overruled by three of late frosts, excessive heat, drought, hail, or savage winds. Months could pass without a cloud in the sky and then, in a fury of thunder and lightning, four or five inches of rain might pound the earth in a matter of minutes, washing away crops starved for even the slightest drop of moisture.

Native grasses and forbs required little maintenance other than a sensible grazing regimen. A dense and intricate root system captured and absorbed the slightest rainfall. During hot, dry weather the opportunistic plants simply shut down growth and grew dormant; when the rains came, the natives renewed their reach for the sky.

Prairie plants cured well when cut at midsummer for winter hay, and some species even held nutritive value when they cured on the stem following frost, allowing for winter grazing. It was a system that supported millions of buffalo, one also well suited for the range cattle that came north with young J.O. and his gypsy cattle drover kind. But the natural scale was vast—the prairie needed every acre of its largeness to succeed. Buffalo moved with the rain and the seasons, covering countless square miles. Even a 3,000-acre pasture, surrounded by barbed wire, was insignificant on such a scale.

Therefore, the ranchers who survived were those capable of amassing vast acreages, who understood that the quality of their grass was the key to success and, maybe more than anything else, who could solicit the support of a friendly banker who would carry a note during lean years and supply cash to buy more livestock when the rains came and cattle prices were high.

J.O. Selman was both a businessman and a plainsman. He grasped the futility of the plow in such fickle, dry country and frowned upon burdening his credit with expensive farm equipment that demanded payment whether crops made or not. Instead he chose to invest in a natural system that not only survived in the harsh climate while supplying his beef cattle with nourishment but, when utilized judiciously, actually thrived.

No doubt the Selmans felt sad when the sodbusters sold out and moved on, because most were good but tragic people, their misplaced common sense overwhelmed by a land hunger that lured them to a region as alien to their Old World agricultural sensibilities as Mars or the moon. Even today you'll find the remnants of their hopes and dreams dug into the bluffs along the Cimarron River, a few weathered boards still covering the gashes in the earth.

These dugout shelters were commonplace living quarters during the rush for land when the Indian territory was thrown open to settlement. Wood was scarce, so the homesteaders simply shoveled a ditch into a hill or bluff, threw a framework roof over the excavation, and then added a layer of sod. Buffalo "chips" and, later, cattle dung supplied fuel for heating and cooking. Frontier cowboy Wiley Cowan once reminisced about a forlorn family of dugout-dwelling sodbusters who approached his own dugout, begging permission to gather the ample manure supplied by Cowan's steers. The family had exhausted the fuel supply in their home area, and faced a life-threatening winter without the only source of heat available on the rapidly depleted frontier.

When one family living in a dugout near the junction of the Cimarron and Buffalo Creek outgrew its quarters, the resourceful settler obtained several wooden barrels and buried them, open end facing the river, in a steep red dirt bank near his primitive home. These barrels became overnight shelter for his children. The remains of the dugout can still be seen, but the rest has gone back to cattle, deer, and the coyotes that prowl the sandy banks of the Cimarron. Those who settled this region were tough, no doubt about it. But there are limits to endurance, even to that of a man so determined to own a piece of land that he would raise his family in barrels entrenched above a riverbank. Within a matter of months or years most drifted back toward the timber and the rain. For a horseback herdsman, their passing was as natural as the wind and the rattlesnakes he'd learned to live with. A herder of grazing animals could form a pact with the whimsical natural world and, with patience, endure. On the other hand, the sodbuster, as soon as his steel plow snapped at the tangle of native grass sod, sealed his own fate as well as the fate of the land, no matter whether he stuck it out a month, a year, or generations.

A homesteader and neighbor of J.O. Selman, this sodbuster kept a scandalous "housekeeper" on the premises.

Over the years the Selman Ranch absorbed a homestead worked by an industrious yet lonely sodbuster, an immigrant who was frugal with his meager earnings and advertised for a mail-order bride.

Eventually the man began to correspond seriously with a young woman from Missouri. The letters that flowed back and forth between them address the woman's determination to keep the long-distance courtship somewhat formal, while the farmer beguiled his prospective bride with speculation concerning the prospects of his wheat crop and a promise of Sunday buggy rides.

The woman sent her suitor the picture he'd requested. It portrays a thin woman, pleasant to look upon, although neither pretty nor plain. She is dressed in the formal attire of the late nineteenth century: large hat, layers of dresses, a new coat she was quite proud of, according to her letters. In the background are the typical trees and rock outcrops of the Ozark region, and it is easy to imagine a young woman desperate to escape the humid, thin-soiled wasteland of one marginal agricultural area for the blistering winds and parched earth of another.

Finally the young woman accepted the sodbuster's written invitation to be his bride, but only after admonishing in print that she looked forward to riding across the prairie with him, but would refrain from the kissing and hugging he'd hinted at until after they were properly engaged.

Judging from old photos of the homestead this man was a diligent worker, and his small farm more than likely was the envy of many in the area. Some remember that he even made a small profit some years, but eventually the drought drove him off the land and his mail-order bride back to Missouri. If this land could talk, it would spill out more songs of heartbreak than all the mournful Appalachian folk ballads combined. For the Cheyennes, living on reservation memories of former glory and the starchy, sugary government-subsidized diet that would lead to rampant diabetes, it must have been retribution in slow motion . . . overdue justice symbolized by a plow blade striking something solid and hard, like buffalo bones.

Cowboys, Coontail Rattlers, and Cottonwood Shade

Reflections of a Ranching Childhood on the Plains

by Sue Selman

Buffalo Creek was a very big part of my life as a child. I remember it as being much different then, wide with beautiful white clean sand and lots of neat rocks and shells. We kids were always attempting to catch minnows by building dams and traps, but we were rarely successful. I would play for hours and hours along the creek, sometimes with a friend, oftentimes alone. We would wade up and down the shallow channel, trying to find a spot that was deep enough for swimming. Mostly we would stretch out flat on our backs or stomachs just to get wet.

~~~~~~~~~~~~~~

We had a rooster on the ranch that seemed big as an ostrich to a little girl. He was the meanest, sneakiest creature, obviously put on Earth just to torment my brother and me. Every time we entered the yard the rooster would sneak up from behind and flog us with his spurs, so we learned to arm ourselves with a stick or a bat. Tom and I complained to no avail, until the morning my mom witnessed an unprovoked attack. She marched out, grabbed the rooster by the neck, whirled him around a few times, and snapped off his head. We had a delightfully delicious chicken stew for supper that night.

~~~~~~~~~~~~~~

I was always determined to find any kittens in our hayloft, no matter how often the mother cat moved her litter. One of the Selman Ranch legends recalls the day my cousin Billie climbed into the loft to discover that a tomcat had killed a new batch of babies. She ran screaming to my mom, who instantly lost her temper, grabbed a butcher knife and a cowboy boot, marched out to the barn, and stuffed that tomcat head first into the boot. A minute later Mr. Tomcat was missing his manhood. Several cowboys were lounging around the barn at the time, and the sight of my mother castrating that cat with a kitchen knife was enough to induce startled looks and a shocked silence. As you might guess, those rough men's respect for my mother suddenly grew to include more than just a fondness for her cooking.

≈≈≈≈≈≈≈≈

Both my brother Tom and I were in the saddle before we could walk. When I was a little older my mother would ask one of the cowboys to bring up old Cactus. She would place me in the saddle, and I would pester and torment that poor old horse for hours. When Cactus had enough, he would head straight for a low branch or the clothesline and attempt to rid himself of his little pest. After being scraped off a few times, I could sense what Cactus was planning and would yell for Mom to come rescue me. Fortunately we always had some wonderful, gentle old horses to ride in relative safety. Several definitely could have been considered our nannies.

Even now I love looking out my window at the horses cavorting in the pasture on a cold winter morning. I also keep some mules, mostly for sentimental value, because when my grandfather J.O. Selman first started this ranch he used mules for almost all the work. There is still some of the old tack left in the top of the barn. Once I asked my dad why we didn't use draft horses, and he said horses couldn't stand the Oklahoma heat. Mules fared much better during our oppressively hot summers. J.O. continued to use mules long after he had trucks, because he felt the trucks couldn't get around in the sand and the mud. I guess our mules provided Harper County's original four-wheel-drive transportation.

As I got older I would ride out to what I felt were my secret spots. I had one special place at a little creek down the road where I spent hours. I also found a beautiful spring at the top of a draw, where the winter grass was like a carpet. I felt that this spot was magical, and it has remained my secret even to this day.

≈≈≈≈≈≈≈≈

The ranch has always been prime rattlesnake territory, and it must have been especially so when Mom and Dad first built this house. Mom never let us kids out to play until she'd circled the yard and completed a snake check. On cool mornings we learned to look closely for snakes before stepping out the door, because the rattlers savored the heat from the concrete walk. Even today we usually find at least one or two rattlesnakes a year in the yard.

It always seemed amazing to me that a rattlesnake could turn a macho cowboy into a squealing sissy. Now and then dad would hire a man who turned out to be terrified of snakes. One summer we had a mouse infestation in the barn and a big bullsnake moved in to take over the job of chief mouser. This would have been fine, except for the fact we had one of those snake-skittish cowboys working for us that year. Everyone in the family knew the snake was in the barn, but to keep the peace we never mentioned it to our new cowhand.

One quiet summer day this cowboy needed something from the barn. Suddenly a piercing scream shattered the silence. Everyone raced to the barn, certain the man was badly hurt. Seconds later the cowhand ran out of the barn, cussing a proverbial blue streak. After he calmed down the cowboy told us he was groping around in the back of a drawer and withdrew a handful of bullsnake. The cowboy swore that from that moment on it was either him or the snake, so later in the day Mom moved the critter to a safer location.

When I first moved back to the ranch after my father died, the place had been abandoned for a few years and the yard was full of snakes. I'd invested in some baby chickens and one day visiting friends asked to see them. The chicks were in a little pen in the corner of the chicken house. I bent down to pick one up and came face to face with a gopher snake with several big lumps in its belly. I knew the snake had to go or I'd soon be out of the chicken business. I tried to grab it, but the snake headed for a hole in the chicken house wall. So I grabbed the tail and we commenced with a tug of war. Gopher snakes are amazingly strong creatures, and I had a hard time pulling it out of that hole. When its head finally emerged from the escape route, I flung it in the opposite direction. Much to my friends' dismay, the snake landed at their feet, and suddenly they were searching for an escape route, too. Later I took the snake, a four footer, to the pasture and let it go free. I think my mom would have been proud.

Every once in a while a snake finds its way into the ranch house. Years ago a stranger knocked on the door, informed us he was lost, and asked to use the phone. Mom pointed to the telephone and then went back to her housework. A few minutes later she decided to check on the stranger and found the phone dangling with someone on the other end of the line yelling, "Hello, are you still there?" But the stranger was nowhere in sight. Sure enough, there was a little snake curled up in the middle of the floor.

Many years ago a family came all the way from Wisconsin to interview for a job. They seemed nice enough, and Mom was pleased with their sincerity. The entire crew seemed excited about the ranch and the job and were willing to accept the position, so my mother asked if they'd like to look around the place. That's when they saw the jars of rattlesnake rattles. Mom explained that we had rattlesnakes, both diamondbacks and the prairie rattler variety. She said the couple turned white, walked out the door, got in their car, and drove back to Wisconsin, never to be heard from again.

During hot, dry weather grasshoppers would carpet our land, devouring everything in their path. During dry years the grass was already in short supply for our cattle, and grasshoppers could deliver a serious economic blow. After they'd eaten all the available vegetation, the bugs would start chewing the paint off the house and the bark from the trees. The yard was ankle deep with them.

Drought meant a continuous shuffling of cattle, and one year my dad accepted a bunch of Brahma steers to feed. These were truly wild cattle, and they acted like they'd never seen people before. We tried to hold them in a corral, but the steers went over the fence like deer. These were mean critters, and smart. The Brahmas would seek out the biggest sunflower patches and hide in the dense cover. When a cowboy rode in after these cantankerous steers they didn't run; they just got ready to fight. After several days of rounding up these wild cattle and then watching them disappear again, Dad finally shipped the entire lot back where they came from.

The fifties were a period of intense drought, and during this time Dad shipped all our cattle to better grass in South Dakota for two summers in a row. The AT&S railroad ran right by our house, and the old corrals and loading dock were still standing. So we loaded our cattle on the train and sent them north to some leased pastureland. Later we drove up to South Dakota to check on our animals. We found plentiful grass with a big river nearby, all near an Indian reservation. One night my dad and the cowboys were sitting around a campfire when the men got an eerie feeling they were being watched. Suddenly several Indians stepped out of the dark. It seems the Indians wanted to trade horses with the cowboys. The Indians' horses were in sorry shape, and they tried all night to bargain. The chief's name was Grass Rope, and he even offered Dad his young redheaded wife for my father's favorite mount.

~~~

I was caught in a prairie fire once when I was a kid and learned firsthand how treacherous these conflagrations can be. My brother and I started the fire accidentally, and when I saw the smoke I jumped out of the jeep Tom was driving and began fighting the flames. It's amazing how sagebrush can explode like a firebomb. I tried to smother some of the escalating inferno with my jacket, but my best efforts were in vain. Suddenly the wind increased and changed directions, and I knew I was in serious danger. The fire began to race across the pasture even faster, and the smoke was overwhelming. I looked around for a safe place to take shelter, but all I could see were miles of tall grass and barbed wire. Then a pickup drove through the smoke and almost ran over me. It was our neighbor, Russell Bradt, and that man literally saved my life. Some 1,200 acres of grass went up in smoke that day, and both Tom and I might have been badly burned if Russell hadn't arrived when he did.

~~~

Cattle pastures are often named for their previous owners, as is the case with the Edwards Pasture on our place. An old homesteader named Moss Edwards once lived in this pasture, in a ramshackle little house with his "housekeeper"—at that time a nice way of saying he was living with his girlfriend. The Edwards pasture has our biggest pond and a swimming dock. This pond provides a great place to cool off after one of our 100-degrees-plus days. On summer Sundays we often gather there with some of our hardworking neighbors and their children for a few hours of fun in the water. My dad caught a seven-pound largemouth in this pond and had it mounted. We also run a trotline there sometimes, because there are plenty of channel catfish in this pond as well.

At one time Dad operated a gypsum mine in the Rock Crusher Pasture on Sleeping Bear Creek. This pasture contains some great quail and deer hunting along with some pretty good fishing, thanks to the beavers and the ponds they've built. It is also a notorious place for rattlesnakes. I believe this is one of our prettier pastures, mostly because of the beauty of Sleeping Bear Creek, plus the surrounding steep red cliffs capped with white gypsum.

The River Pasture originally encompassed 3,000 acres, but lately I added cross fencing for the benefit of pasture rotation. I had mixed feelings about cross fencing that pasture. I'm sure it was in the best interest of the grass and the wildlife, because cattle prefer certain spots within a pasture and they can decimate the grass if they stay in one area too long. Now I have more gates, two new water gaps, and five more miles of fence to keep up. I often wonder if my parents would have ever cross fenced the pasture. Dad loved to watch the prairie chickens and was an avid quail hunter, but like most ranchers he wasn't aware of the habitat requirements they needed to survive. Both Sleeping Bear Creek and Buffalo Creek run through this big chunk of grassland, so the terrain tends to be rough. It was always a big job to gather cattle out of this pasture, a task requiring several cowboys.

The pasture we call the Narrow Neck borders the Cimarron River. It is a big pasture, around 2,300 acres. If you look east from our ranch house, in the distance you'll notice what appears to be a very narrow opening between two large cliffs at the point where Buffalo Creek merges with the Cimarron, thus the Narrow Neck. This is the roughest pasture to work cattle in, due to numerous steep canyons, brushy draws, and ample rattlesnakes. The old loading pens next to the abandoned railroad track are still standing, along with our corrals. When we worked cattle there, Mom would fix lunch and pack it in the trunk of our car. We'd eat in the shade of a big cottonwood tree. This pasture has some great scenery, due to the big vistas that include the Cimarron, the bluffs that border the river, and a several-thousand-acre salt plain. There are some huge old buffalo wallows in this pasture, plus the remains of our very own ghost town, Salt Springs, site of the last Old West–style bank robbery in Oklahoma, including an outlaw on horseback and a getaway stymied by a bottle of whiskey.

The Dorman Pasture is due west of Narrow Neck. Legend has it that Mr. Dorman made moonshine whiskey here. My son found an old crock whiskey jug while hunting here, so maybe the stories are true. This pasture contains several good springs, and my dad converted three of them into ponds. Several of the ponds don't hold water all that well, thus our attempt at stocking fish hasn't fared too well, either. Still, there are lots of bullfrogs in these little prairie reservoirs, along with plenty of waterfowl in the fall.

Bordering the Dorman Pasture going west is the Stud Pasture, where my dad kept his mares and stallion when we raised our own horses. The upper pond in this pasture is the first one my father built. It is a beautiful place, with clear water surrounded by white gypsum cliffs. This pond acquired a slow leak, so Dad built another one directly below the original dam. We kids loved it when we had a new pond to swim in, due to a lack of moss clogging the water and the absence of stickers along the shore. A new pond was as close to a fine ocean beach as we'd ever get.

Both the North and Far North Pastures contain little spring-fed creeks that feed nice ponds. The one in the Far North Pasture was the last pond built on the ranch and is called Sue's Pond in honor of yours truly. When my youngest son was just a baby I arose early one morning, grabbed my fishing pole and my child, and went fishing at my namesake pond. I sat on the dam with my son in my lap and caught fish after fish, eventually bringing home a full stringer. No one had bothered to fish there, since the pond was new, and they were surprised at my success . . . and maybe a little miffed that I beat them to that first big catch.

The excitement surrounding roundup remains unforgettable. It involved meeting new people, a unique style of camaraderie and seeing the pleasure in my parents' faces as they watched their cattle come off the range. Plus it was a change from the monotony of our daily routine. I remember pestering my mom and dad to let me help work the cattle. At first I was puzzled by the lack of enthusiasm on their part, since we were all expected to pitch in and help with just about every task on the ranch. Mom tried to explain that I was too young, that it was a long day in the saddle and no one had time to take care of me if I got hot, tired, and thirsty. Still I persisted and finally wore my mom down. She loved to ride and had helped work cattle when she was younger. She might not have said so, but she knew exactly how I felt.

I must have been around eight or nine years old when I went on my first roundup. I was given an older, trustworthy horse to ride, and I was wild with anticipation. I gathered up my clothes at bedtime so I could dress in a hurry the next morning, polished my boots, saddle soaped my saddle and tack. I wanted to prove I could be ready, as my mother had voiced her concern that I wouldn't be able to get up in time, considering how early the men started for the pastures. The horses were fed in the evening and kept in the corral overnight. They would have to be saddled before the sun came up. All the horses were fed together, then the ones to be ridden would be caught and tied to the fence to be saddled. The rest were turned out to pasture. I knew if I didn't get up and out to the corral on time there was a good chance that some cowboy would let my horse go free and I'd be left on foot . . . and miss my first roundup.

When I was a kid, the area ranchers "neighbored," or helped each other work cattle. Often in the mornings before we worked our calves, I would awaken long before dawn, hoping it would soon be time to get dressed for breakfast. Then the rumble of a truck and trailer or the bright glare of headlights coming down the road would propel me out of bed even before the birds started singing. Long before the sun came up our yard would be full of pickups, trailers, and hopefully, some new cowboy faces.

Eventually I turned out to be a top hand. Those mornings on horseback taught me to appreciate getting up early. I wish I could adequately describe watching a sunrise on horseback, the smell of dew on sagebrush, the incredible rainbow display of colors in the sky, the mist hovering over the river bottom and the way the low, clinging fog turns from gray to pink to purple. Even today I remain in awe of the brilliance of the sun as it first cracks over distant hills where red earth, white gypsum caprock, and green shades of the grass and sage all seem to merge. It can cause even the crustiest old cowboy to brim up with a tiny tear to see such beauty.

Gathering cattle from the distant pastures demanded an early start, because we usually trailered our horses to these outlying areas rather than riding there from the ranch house. Saddling up and heading out before daybreak was always a tense time for me, because I'd watched more than one belligerent horse test its rider during a little early morning bucking session. Even the horses that didn't buck were excited and full of energy, given to prancing, pawing, and side-stepping. I was always terrified that I would get bucked off in front of the veteran cowboys, but fortunately that never happened.

When we reached the pasture, the cowboys would form a circle and my dad would pair us up and assign each pair a certain section of the pasture. Then he'd explain how he wanted the pasture gathered and determine what time he expected us to have the herd gathered and where he wanted the animals held. Dad didn't allow his cattle to be herded at a run or roped unless it was absolutely necessary.

Some pastures were easy to work while others could be tough. The River Pasture and the Narrow Neck were the biggest and also the most difficult to gather. Both pastures contain a number of narrow draws filled with plum thickets and sumac tangles—perfect places for a cow to hide. Generally our cattle were fairly easy to gather due to the excellent riders who turned out to help us. I learned the art of gathering and herding over the years under the tutelage of many a great cowboy.

≈≈≈≈≈≈≈≈≈≈≈≈

I acquired a wonderful cow horse when I was about 13 years old. He was a big sorrel with a little bit of white on his face and white stocking feet. This horse was 16 or 17 hands tall, and it took me a while to figure out how to get in the saddle. This was a veteran, no-nonsense horse, and I'd been accustomed to easygoing kid horses that I could clown around on. The first day I rode my new horse I casually flipped my foot out of a stirrup and was going to throw my leg up over the saddle horn and relax. Well, that big sorrel came unglued and just about ditched me. I was riding with Billy Joe Caldwell that day, and he said, "Oh yeah, Sue, I forgot to tell you. Don't try that kid stuff on that horse."

As I got to know my new sorrel, I soon realized that he was one hell of a horse. So did everyone else, for that matter, and I became the envy of many of the cowboys. I found out his name was One Way, and there were several stories concerning how he got his name. One was that the horse was so big and powerful that he looked like he could pull one of those huge old one-way plows. The other was that when the men were breaking him, the sorrel only wanted to go in one direction, which was back to the barn.

One Way was born to work cattle. He instinctively knew exactly what he was supposed to do. He would watch the cows closely and always seemed to know when one was ready to make a break and escape. One Way was so quick and intuitive that I had to be ready at all times or that horse would leave me in the dirt. If a calf was lagging behind, One Way would reach down with his nose and gently nudge it along. I'd give just about anything to find another horse like him. One Way lived to be about 25 and was beautiful throughout all his years.

≈≈≈≈≈≈≈≈≈≈≈≈

My family has owned several legendary horses. Back in the open range days, my grandfather John was out somewhere in the wild country between the North Canadian and Cimarron Rivers with a big herd. After being by himself for a while he decided to ride back into town to visit his parents. He kept the visit short since he was needed on the range, and after a few hours he started back toward a cow camp at Fort Supply, where he hoped to obtain a fresh mount. The cowboys in camp said they were out of fresh horses, so John had no choice other than to stick with Bay Ed, a tough old cow horse he often rode.

Bad news had spread through the cow camp at Fort Supply. The daughter of a cowboy working out of the Packsaddle camp was seriously ill. John knew he needed to get the message to the cowboy, who was also a close friend. So he and Bay Ed started down the long trail to Packsaddle. When John reached the Packsaddle cow camp, he relayed the message and asked for a new mount. But once again, the men proclaimed they were fresh out of horses. So John was forced to ride Bay Ed all the way back to Woodward.

Finally, after another long journey, John and Bay Ed were trudging down the muddy main street of the bustling frontier cowtown. Suddenly a wind-blown piece of paper came flying down the street and the "exhausted" Bay Ed started bucking like a colt. John said he decided that if the damned old horse had that much energy left, he'd just ride him another 30 or 40 miles back to Alva.

When the cowboys worked cattle, the womenfolk and kids would wait anxiously for the herd to come in sight. Sometimes we'd stare out of our big picture window all morning. Often we'd hear the cattle coming long before we could see them. The herd would emerge from a cloud of dust, and I loved to hear the cattle bawling and the cowboys calling out to keep them moving. If we were lucky, the riders would hold the herd in a pasture near the ranch house where they would cut out the bull calves before they were driven into the corrals. It was a wonderful sight, watching those horses work so hard at cutting out those cattle. When the herd was corralled, a flurry of activity followed. Water jugs, medicine, and supplies suddenly appeared. Someone would fire up the branding-iron heater, and we'd listen for the loud roar of the flaming gas. Cows and calves would be bellowing so loudly it was hard to talk above all the noise. Even so, it was music to my ears . . . real cowboy music, then and now.

We raised pure Hereford beef cattle, registered and grade. The grade brand is a "T fork" (a T over an upside-down V), and the registered brand is a wine glass (upside-down T fork). Dad would stand in the corral with an old cane and pick out the registered calves so the cowboys could separate them. This ritual took a great deal of concentration on Dad's part, along with plenty of practiced skill. And he certainly didn't like to be distracted. Occasionally someone might dispute his choices—but not often. The calves were driven into a holding corral, then into the chute corral. I always felt sorry for the cowboys who faced the task of pushing those calves out of the little corral into the chute runway. If they didn't get kicked they sure got a slick green "white-washing." I was always told that calves can't kick very hard if you push right up against their hindquarters. Unfortunately I never got up the nerve to test that theory.

Once the chute was full we would slide a post in place to keep the animals from backing up. I usually inherited this job and did it carefully and well, since there was little room for error. After hot shots (electric cattle prods) came along, they helped speed up this process somewhat, as the calves didn't much care for that serious jolt of electricity. Occasionally someone would get a little heavy handed with the hot shot and the animal being punished would seem to go crazy. This brought a serious reprimand from my dad, who had little time for any sort of nonsense when we were working cattle.

Operating the chute gates can be something of an art, requiring practice and concentration to get it right. Once a calf was in the chute, a flurry of activity followed. Ears were notched, shots given, and of course branding, all accompanied by the odor of burning hair and the bellowing of upset calves. Registered bull calves were spared the knife. Otherwise, the males were castrated.

My dad usually did the cutting. By the end of the day we would have a big bucket filled with calf fries (testicles), a true cowboy delicacy. One of my favorite jokes was to inform any visiting city slicker that we were having rocky mountain oysters for dinner . . . then I'd wait until the stranger had a mouth filled with this peculiar appetizer before explaining exactly what they were eating. Some took it well, others didn't. However, my mom didn't care. She liked calf fries so much she'd stand at the sink and clean them for hours.

When we worked our registered calves, Mom would come out with a tablet and keep a written record of each calf, heifer or bull, issuing each a number so she could send off for registration papers. The registered cattle kept their horns, one of which received a branded I.D. number. This corresponded with a tattoo placed in the ear. Unregistered or grade calves were dehorned.

My dad preferred to work his calves later than most, closer to the heat of summer. We'd try to finish before the hottest part of the day, but it wasn't always possible. Cows separated from their calves have good reason to be upset, and now and then some hot, mad old cow would go crazy and come over the corral fence in her fury. Cowboys can run for safety like jackrabbits when a wild-eyed, slobbering, bellowing cow with murder on her mind decides it's time for them to die. Grandfather Cowan always helped us work our calves, but unfortunately he wasn't very light on his feet. There was a big, tall feed trough in the middle of the corral, and when he knew he couldn't clear the fence, my grandfather sought refuge in that trough. I'm sure it saved his life on more than one occasion.

Once the calves were worked we would cut out any cows that needed attention. Then it would be their turn to take a run through the chute. Obviously the grown cows were a lot tougher to work than the calves. Then we'd hook up our mechanical rig and spray the herd for flies. Then a few cowboys would move the herd back to pasture and watch them until the cows were "mothered up" with their calves. The process of working our cattle could go on for several days.

≈≈≈≈≈≈≈≈≈≈≈≈

All the cowboys knew they were in for a treat when Wilma Selman was cooking. I'm sure my mom's reputation for laying out a lavish table was incentive enough for many of them to help us work cattle.

Mom married my dad when she was 16, and she didn't know much about cooking or anything else domestic, for that matter. But as a rancher's wife, she had to learn fast. Soon after they married, Mom found out it was her turn to cook for the cowboys working at the ranch where they first lived, the Selman place at Fort Supply.

Mom immediately panicked, because she didn't know what to cook. But J.O. came to the rescue and told her to prepare a big roast. He selected the meat for her, placed it in a pan, and smeared on about a pound of butter before the meat went in the oven. With J.O.'s help, Mom's first cowboy meal was a hit. I still have two of the little wood-burning stoves she used, and when I look at them it amazes me that she could do so much work with such simple tools.

On cold winter nights Mom would heat up bricks on the stove, wrap them in rags, and then stash the bricks at the foot of the bed to keep her and Dad's feet warm. Before going to bed she gathered her houseplants on a table, placed a burning lantern in the middle of the plants, and then covered the entire collection with a blanket. Hopefully the blanket would trap the heat from the lantern and keep her precious plants from freezing, maybe even without starting a house fire. In the mornings she'd chip the ice out of her fish bowl. Dad recalled that as newlyweds, they ate prairie chickens and quail a good deal of the time. Mom and Dad eloped to get married, and their parents weren't particularly happy about it . . . certainly not happy enough to spoil them.

Mom's day in the kitchen started at 4:30 in the morning. I can still smell the bacon and the coffee; it was a wonderfully pleasant way to wake up. The cowboys were at the table before daylight, ready to eat. Most were just good, hardworking boys who thoroughly appreciated a decent meal. And some had no bottom to their appetites.

Mom tried to be frugal with what money we had, and feeding cowboys can be expensive. She had to learn to fix a reasonable amount of food and then stop; otherwise, some of them would still be eating. I was always glad to have the cowboys at the table, because they entertained me and I tried my best to return the favor. Most of the men had great manners and were very considerate, but every now and then we'd find an exception like Slurpy, who slurped his coffee and soup and generally made unpleasant noises at the table.

A typical meal on days we worked cattle would consist of fried chicken with mashed potatoes and gravy, or a huge roast that looked like a mountain of meat. Mom would add green beans with bacon, onion, and potatoes, cornbread or biscuits, and fried potatoes. She topped this off with a double-layer chocolate cake, or cherry, apple, or gooseberry pie.

When the food was ready, Mom would ring the dinner bell. This was the old Kirby School bell, where my mom attended classes as a little girl. It looks like a smaller version of the Liberty Bell, and on a calm night you can hear it ringing from a couple of miles away. Shortly after the bell rang, the cowboys began to materialize around the table. I can still hear them laughing, talking, and stamping their boots to clean them as they came up to the door, spurs jingling. These hot, tired men entered a cool house with a table piled high with wonderful homemade food and a big glass of iced tea. It must have seemed like heaven.

Clay Heabal was an older cowboy who lived in our well house. He was my buddy. In the evenings Clay and I would sit on an old bench in front of the well house and talk like a couple of old troopers. I think I liked him best because Clay would actually carry on an adult conversation with me.

Clay cowboyed for us during the early fifties, and during that time someone lost or abandoned a little cocker spaniel out on the highway. Clay adopted that dog and called her Poochy. Later Poochy adopted a litter of orphaned kittens. Mom and Grandma didn't really approve of all the time I spent with Clay, because he liked to tip the bottle.

≈≈≈≈≈≈≈≈≈≈

One of my favorite cowboys was a man named Billy Joe Caldwell. He worked for us for several years and was always a good friend who taught me a lot about horses and riding.

Billy Joe was in his late twenties and a good-looking man who loved to go to town and have a good time. One morning when Billy still hadn't made it back home from a night of carousing, my cousin and I eased into the bunkhouse and left a dead snake in his bed and a coyote tail tied to the doorknob. Billy made it home shortly after sunup, and my cousin and I were hiding nearby when the cussing erupted.

I cried for several hours when Billy Joe left the ranch. We spent a lot of time together in the saddle and he taught me the things every cowgirl needs to know—where and how to cross a creek, when my horse was getting too hot, how to gather cattle. Billy Joe had a great sense of humor and was always telling jokes. I doubt that he ever knew how much I missed him.

≈≈≈≈≈≈≈≈≈≈

It was always hard to find good cowboys and then even harder to get them to stay for very long. The pay wasn't too good and the work was always rugged. Once Mom hired someone through the local employment agency and then drove to town to pick him up, since he didn't have a car. She was pleased with the work the man did and relieved that they had finally found a good hand. But then, after a few days, the sheriff drove up and arrested our top hand for bank robbery.

Buffalo Creek Sketches

A Few Words for and from the Land

by Gary Lantz

In late February the Selman Ranch is as barren as tundra, the colors subtle like those found in subdued Southwestern art, or maybe in the way that Navajos use earth-tone dyes to produce hand-woven rugs that collectors cherish.

This afternoon the sparse colors of winter are magnified by an eerie light preceding an approaching thunderstorm. Everything seems strangely illuminated. The pale gray leaves of sagebrush contrast orange clumps of range grasses, all accented by the sharply pointed, upright fleshy green leaves of soapweed yucca. Outlines are sharply etched, and for a few moments an otherwise rather bleak winter landscape seems to radiate light. Then clouds collapse upon the sun and details disappear back into shadows.

The approaching clouds appear bruised and angry, and although it is still early in the season the time for threatening skies is already upon the southern plains. For several days now south winds have sucked moisture up from the Gulf of Mexico, and afternoon temperatures have warmed into the high sixties. This rapidly advancing cold front will produce a clash of contrasting air masses that sparks the spring rains as well as hailstones and tornadoes. Yet as threatening as they can be, the banks of low clouds piling up on the horizon promise rain. Rain is holy here and the clouds symbolize a benevolent God, even if barns blow down and trees are uprooted at His random touch.

Just before the rain tumbles down in torrents, the bare limbs of the cottonwoods along Buffalo Creek seem to pick up an electric charge from a momentary flash of setting sun. They shine a fine, gleaming silver, then appear to catch fire. In a matter of seconds the storm swallows the sun once again and light fades on the silhouette of a bobcat emerging from a plum thicket, winter coat in style with traditional ranch colors like sorrel and buckskin. The cat slips back into the purple tangle of a sand plum thicket, and there is nothing left to do but hope for an encore, await the next rumble of thunder, and debate the possibilities of this spring rain reverting to an immobilizing winter snow.

March on the southern plains means wind. Life on the plains is orchestrated by seemingly constant winds, and if you cannot tolerate the wind, this land will make you crazy.

In March sometimes the winds truly seem to go mad, and plains dwellers with a history here cannot help but recall the Dust Bowl of the Depression-era 1930s. Occasionally the cotton farms in southwest Oklahoma and the Texas Panhandle rise up to remind them, and although the dust clouds are generally of short duration these days, they can be intense.

But along the banks of Buffalo Creek the soil is knitted with the tight stitch of native grass, forbs, and thickets of native shrubs. The wind can howl here, but it won't lift the land and redeposit topsoil somewhere else. Mostly the wind is nothing more than cause to be a bit cranky when strong southwest gales return to quickly parch a recent gift of summer rain, or a cold north banshee of a blow knifes through clothing following a series of balmy winter days.

This particular March morning dawns with a hint of spring, so we've walked the bluffs overlooking the Cimarron River under a 65-degree sun, watching mallards in breeding plumage lift from the silver ribbons of the braided channels far below.

The sun is so bright that the vast salt flats of the Cimarron are almost blinding, like mountain snow. Everything seems to sparkle: the salt, the meandering shallow arteries of the river, the layer of gypsum beneath our feet that caps the red clay bluffs like a crown.

Crystal selenite gleams where erosion has exposed it, and I wish for a pair of sunglasses so that I can delineate objects that move along the edge of the river, small animals shuffling from shadow to shadow. We linger to explore the remains of a homesteader's dugout, the family probably gone for at least a hundred years. A few graying old boards still form an open roof, merging with a wooden frame set within a ditch dug back into the red clay bank.

Years ago the roof would have been covered with sod, and at night rattlesnakes would have come calling. Rather than enlarge his earthen shelter, local legend has it that this sodbuster sank wooden barrels into the red bluffs bordering the river to provide sleeping quarters for his children. The skeleton of a roof we can still see today represents how the hopes of one homesteader played out. Maybe the epitaph should read, here lies the dream of a farmer, living in a ditch beside a river of salt, drinking gyp water that tied his guts in a knot, praying for rain to nourish the seed planted in dirt meant to support only native grass and buffalo.

Sadly enough, this family's life represented the norm for immigrants flooding into northwest Oklahoma when Indian lands here were opened to white settlement. Most homesteaders met with a bleak and meager way of life, drawn by illusion to face the reality of heat, drought, blizzards, tornadoes, wildfires, rattlesnakes, grasshoppers, bad water, lack of fuel, and a grazing culture already in place that didn't want them, openly despised many of them, and mostly didn't give a damn if they survived or not.

The cattlemen either waited the sodbusters out or, more than likely, gave some a not-so-gentle nudge in some other direction. In reality there were too many would-be farmers in a region poorly designed to hold any at all. Probably the nudge, no matter how coldly administered, was a merciful one.

This Cimarron River dugout settler who dreamed of growing dryland wheat and corn along the banks of a salt river must have lived in his imagination while he worked the raw earth. Today cattle graze around the meager remnants of his life here, these weathered boards and a gash in the ground grown up in vines and weeds, all a testimonial to the futility of farming a land of fickle rainfall. Eventually we leave the homestead behind and follow a deeply worn trail up over the bluff and onto a wide tabletop meadow sloping slightly south, toward the valley of Buffalo Creek and away from the red dirt epitaph of self-induced delusion.

In the middle of this upland meadow awaits an oval depression, mud-encrusted and dry at this particular time yet as big as many manmade stock ponds in the area. It's one of the largest buffalo wallows I've ever discovered, maybe a hundred feet across. Now that I've seen them spiral away from the wallow like spokes on a wheel, there is little doubt that the deeply worn paths leading up from the salt plain and Cimarron valley are old buffalo trails, still tended by cattle, deer, and an occasional pedestrian writer whose daydreams include both massive bison herds and Cheyenne hunters . . . with little room left over for hardscrabble settlers who store their children in barrels.

≈≈≈≈≈≈≈≈≈≈

The road to Salt Springs is little more than a pair of pickup tracks in loose sand, busting through thickets of plum and skunkbrush sumac where the Buffalo Creek valley narrows just before dumping into a delta of gleaming white salt and the Cimarron.

Along the way we pass the carcasses of freshly killed coyotes, compliments of the federal trapper, your tax dollars at work. Seven in all, apparently healthy judging from size and pelt, each probably a villain due to reputation more than deed, but dead anyway.

Killing coyotes provides a few government-subsidized jobs scattered across our western range country, and also sport in the form of hunting with both rifle and dogs. The one thing killing coyotes doesn't seem to do is drive this adaptable little predator toward the endangered species list. Coyotes appear to be expanding their range and numbers, while the future of other predators like the swift fox is more precarious. The swift fox is attuned to a specific niche within its historic environment, while the adaptable coyote has learned how to rocket out of brushland bordering Beverly Hills and snatch a poodle snoozing by some movie star's swimming pool. Plenty of people hate the coyote, leaving me room to admire the animal's grit in the face of an ongoing desire to eradicate it. Besides, I'd miss the evening serenade if the trappers were too successful, all that high-pitched yipping that mixes so well with panoramic prairie sunsets. It appears that the war on coyotes has only made them smarter. In fact coyotes remind me of the people who have managed to stick it out on the plains. Both are tough, resourceful, and prone to let off a little steam when it's time to howl.

Salt Springs is now in sight, or what is left of it. Leftovers include a pair of red brick silos plus cattle shipping pens perched beside the old railroad right-of-way, now sans tracks. The rails were removed years ago, but the pens are still stout enough to hold a few hundred wild-eyed Texas steers.

Selman heirs have a painting of the old Salt Springs, done by a local artist on commission. Those who could remember told the artist about certain scenes that lived on in their minds, and he began to paint. According to these oldtimers, the finished canvas is accurate and depicts cowboys unloading cattle near a small, neat little western community with a bank and two fine new brick silos.

Today the area is grown up in plum thickets, and few remember that Salt Springs was the scene of the last horseback bank robbery in Oklahoma. Of all the legends based upon this deed, the one I like best also seems the most likely. According to a Selman neighbor, some cowboys got one of their number drunk and dared him to rob the bank. The young cowhand miraculously managed to pull it off, escaped on horseback, and then promptly passed out. And that's how the mounted lawmen finally found him.

Sleeping Bear Creek empties into Buffalo Creek from the south and drains a picturesque valley bordered by red bluffs capped with white gypsum. This little stream is one of three that forms the heart of Selman country. Old J.O. knew from the start that he wanted to own land drained by Buffalo, Sleeping Bear, and Sand Creeks. Son Bob kept the core lands around these three drainages in the family, and the modern ranch remains rich in wildlife because he had the foresight to do so.

Maybe the most spectacular feature of Sleeping Bear Creek is the number of clear, cold springs that feed the constant flow. A good horse can easily jump this narrow little waterway, except where Bob Selman has 'dozed out a pond or where beavers have made numerous ponds of their own. What matters most is the stream stays up and running in the blistering heat of July and August, and that the springs still flow even in moderately severe droughts.

There is never enough water on these prairie/plains. Never enough rain, never enough springs, never enough drinkable, mineral-free water for gardens, for grass, to rob the cottony dryness from one's thirst. So when you look on a map and ascertain that the Selman Ranch appears to be where the best of the creeks come together, you can't help but wonder if J.O. wasn't the luckiest cattle drover ever to leave the state of Texas, or if his hard work didn't simply earn the respect of bossman Tandy, who obviously had plenty of connections and a strong line of credit at the bank.

It's a warm afternoon in early March, and a pair of killdeer are mating near a pair of ponds, one on Sleeping Bear Creek and the other damming a spring that feeds it.

It's nice to see that the little brown and white plovers have something on their minds other than shrieking their dismay at my intrusion. Allegedly the birds sing a phrase that sounds like "kill deer," thus the name. However, the call I generally hear sounds more like a midnight brawl at a tavern on the wrong side of the tracks, with an aggrieved girlfriend screaming at a pitch near the cusp of my range of hearing. Killdeer, like marauding crows and bluejays, can test the patience of even the most forgiving naturalist when they tune up to warn every living thing in the region of one's approach.

Dull brown stalks of cattails, thickly grown and head high, indicate the origin of the spring just upslope from the stream channel. Nowhere is Sleeping Bear more than about 20 feet across, running strong with clear water over fine sand. Bright green strands of filamentous algae sway in the current, and the shells of an invader, a small yet rapidly colonizing Asian clam, litter the area. Another Old World intruder, the equally invasive salt cedar or tamarisk, has usurped the domain of natives like cottonwood and willow, both good for wildlife while the newcomer is not.

Most of the native trees that once shaded Sleeping Bear Creek have long since returned to the soil. Those that are left are very old and grow well back from the stream bank, back where some 50- or 100-year flood managed to moisten a fertile seed bed years ago.

Today the phalanx of salt cedar, in places an impenetrable tangle, resists the incursion of native seedlings and also sucks hard at the water table, keeping exploratory native roots from reaching moisture reserves in the sand. Too many of the remaining ancient cottonwoods occur along streams infested with salt cedar, and these spreading monarchs are generally well back from the stream bank. Determine the age of these survivors and you may have the date when extreme flash flooding last swept down these valleys. A glance at any one of the gnarled, weather-pruned giants seems to reveal a tree that has served as a great blue heron nesting platform or turkey roost for at least a hundred years. My guess is that the flood that carried the cottonwood seeds that resulted in these trees was of legendary proportions, though any witnesses have since gone to dust. It might even rival the relentless droughts of this region in dramatic recollection, although any form of moisture here is welcome, even floods that tear out fenceline watergaps and collapse bridges.

Therefore, we often stop to admire the solitary cottonwood or black willow, main trunks as broad as a whiskey keg and sometimes not much taller, limbs broadly spreading, the signature of a tree grown without restriction to the sun. Each provides meager shade and a perch for owls, crows, flycatchers. Someday we'll know how to efficiently eradicate invaders like salt cedar. Until then these stream banks will remain unfulfilled, incapable of living up to their wildlife habitat potential.

Fortunately the wildlife habitat survives mostly intact farther up the Buffalo Creek tributaries, especially in the ravines cleft in stark buttes overlooking the Sleeping Bear Creek valley. Many of these contain short, stout elms, now with swollen flower buds, the trunks emerging from a snarl of sand plum and skunkbrush. Today these narrow little canyons are alive with hundreds of robins and mourning doves. A nice young buck, a second year six pointer, explodes from a sumac tangle. Meadowlarks and bluebirds are both vocal and common. Each song is the voice of hope, even though more freezing winter weather could spill over the horizon in a heartbeat. The birds, hormones surging to the siren song of lengthening days, have little patience with negative thinking.

~~~~~~~~~~~~~~

Within days the jet stream dips, the weather changes, and a bitter wind coats the countryside with ice. Fortunately winter's encore at this latitude rarely lasts long, and within a couple of days the wind has died back and the temperature has climbed into the forties. I listen for birdsong but the prairie has been invaded by a new music, the sound of ice melting, a metronome of constant dripping.

Patches of snow on the distant hillsides grow more ragged with each sunrise, and finally we are left with water standing in puddles or broad shallow lakes and plenty of mud. The roads are practically impassable, and flooded farm fields summon migrating ducks.

A walk down the banks of Buffalo Creek on the first warm evening following the onslaught of sleet and snow is a lesson in what deer eat and how much byproduct they can generate in the process. The path is littered with deer droppings, the brown ovals about the size and color of chocolate drops.

Every few steps we stop to investigate a clump of sand sagebrush. It appears that the plants have been browsed by deer, judging from the way the stems have been cropped, tiny strips of bark loose and dangling. Soapweed leaves, evergreen and fleshy, have been nibbled on, too. Beneath these plants the earth shows a faint verdant blush as early annual grasses grow on the trampled banks, concentrating grazers eager for the taste of the first fresh greens of the season.

Just before sunset a great blue heron sails overhead on slow wingbeats, a modern-day pterodactyl, the bird's coarse squawk not far removed from a dinosaur's cough. The solitary giant, largest of the native waders, leans into its feathered oars and disappears downstream, out of sight over a distance of several miles due to deceptive speed. A coyote chorus welcomes the winged visitor while the sky seeps blood where distant hills meet metal-blue heavens. We stop to stare at the sunset, and suddenly the entire western sky turns crimson, bright as a bonfire, totally overpowering for the few seconds it lasts. For a moment it's easy to envision Coronado and his bewildered expedition of explorers, thirsting for gold when they left the New World Spanish settlements but now
~~~~~~~~~~~~~~

thirsting for water and wandering this vast universe of grass, stopping for a moment to be mesmerized by just such a sunset, the obligatory priest, muttering "Sangre de Cristo," and making the sign of the cross as day passes over into a night of prowling wolves and equally curious Indians.

A hard freeze overnight helps with the problem of the mud, but it's bound not to last as the sun seems particularly focused on the plains today. For a few minutes at least we can roam the ranch house premises without losing a boot in the muck while admiring the way the old Siberian elms accommodate a flock of loud robins and a scattering of equally vocal bluebirds.

Deer have visited in the night, and their tracks and scat circle the house and outbuildings. Exit tracks lead straight to a ridge of white gypsum just beyond the bunkhouse, and we follow.

The tracks climb through sagebrush, clumps of bluestem grass, plum thickets. Low-growing elms are laden with red-brown flower buds, an early spring food source for some birds. We find where deer have been browsing sumac and dislodge loose rock while we climb. Chunks of white gypsum rattle down steep slopes, skipping over equally white patches of sleet that persist in the shade.

Soon an interloper enters the picture. In this instance it's a rather porcine striped skunk, out on a daylight mission known only to the odiferous traveler, who soon encounters the ranch mascot, a basset hound with ears like an elephant, very little ground clearance, a voice of thunder, and the brain of one of the resident land snails.

Buford the Basset challenges the skunk and gets a light misting for his efforts, just enough to stop the hound in his tracks and start the self-cleansing process, this time extremely loud and rather arduous licking. Tonight Buford will not be welcome in the guest house, no matter how hard he begs.

The first week of April arrives at the Selman Ranch warm and windy, a moisture laden wind blowing up from the south, the temperature this afternoon in the seventies.

Elm trees scattered along the Cimarron wear a dull, pastel tint of green, not from new leaves but the green of tiny oval seeds soon to mature. A porcupine forages in the crown of one of these tough little trees. We watch from atop one of the riverside bluffs, and it's difficult to tell if the bizarre-looking creature is feeding on seeds, the tips of twigs, or both. Whatever the menu, the porcupine is dining in a businesslike manner, long yellow hairs blowing in the wind, the routine broken only by a break to scratch whatever parasites plague the skin protected by those painful barbed quills.

Tiny, biting gnats are swarming this morning, and keeping them out of eyes and nose is a necessity until the sun climbs high enough to summon the wind. The insects make it difficult to concentrate, and we walk toward a gypsum outcrop overlooking the confluence of Buffalo Creek and the Cimarron. Miniature bright yellow flowers, an early member of the mustard family called bladderpod, are in bloom, and last year's tussocks of dried bluestem grass are green at the base with ample new growth.

The cottonwoods along Buffalo Creek are heavy with flowers, those dangling little knotted ropes that twist in the wind and swell with green pods containing white, cottony seeds at maturity. Secluded canyons amid the river bluffs harbor sand plum thickets covered with dense white flowers that look like patches of snow in the distance. In places the plums mingle with equally impenetrable clumps of wild currant, the long, tubular, red and yellow flowers contrasting the sober blossoms on the plums.

These are the "plains" that so many Americans stereotype as monotonous, but I feel like I'm walking in the middle of a freshly painted canvas by Georgia O'Keefe. The white caprock, red dirt, plum blossoms like snowflakes, the butter yellow and crimson of the currant and subtle shades of green mingle and merge delicately, maybe too delicately for those who speed past in search of a picture postcard moment in distant mountains.

Late in the evening we gather on a point of rock overlooking the Cimarron. Coyotes are "howling" in the thickets along the river channel, and I reflect on the clichés of wild canine vocals and decide that howl may fit a wolf, but coyotes are as shrill as any Irish tenor when they yip, bark, and yodel.

Just as the coyote chorus subsides, three deer suddenly appear below us on a spit of sand mid-river. In a moment another whitetail joins these three, bounding and splashing through the shallow, loosely laced channels.

With the sun low in the sky the river shines like a newly minted silver coin and merges with the sheen of the vast salt flats in the distance. The quiet of this place at this time of day is almost stupefying, until the chortle of wild mallards and the coltish whinny of a greater yellowlegs rises from the river like an audible mist. Binoculars reveal the long-legged shorebird in flight over shallow water surrounding a long, narrow sandbar. Soon the streamlined gray migrant lands and begins to feed with quick, almost exaggerated steps, like a circus performer walking on stilts. The head bobs and the long bill probes the shallow water with a sewing machine motion, gleaning crustaceans, worms, or maybe minnows from the living river, a wild river in its own right, a river that few people truly know anymore and even fewer understand.

We are deep into the evening, driving out of the river bottom and its plum thickets, and deer burst out of the brush and disappear over the bluffs by finding a crease and following ancient trails. A kingfisher indicates its displeasure at our approach with a loud, hoarse chatter. The sound ricochets down from a chunk of white gypsum protruding from one of the bluffs, the rock blood red as a waning sunset catches the uppermost edge and seems to set it afire. I cannot help but feel that somewhere nearby a big cat is stretching its long, lithe legs and preparing for the evening's walkabout, dreaming of fresh venison for a midnight snack. Biologists say cougars may enter here only as an occasional migrant, but I believe that the deer population here warrants a full-time feline resident with a taste for wild meat and the strength and cunning to kill it here in the sand. The bears are gone from the region, and the river is lessened by their absence. Anyway, I hope a big cat is here somewhere in the buttes and bluffs and canyons, surveying the night by the light of the moon and leaving round, undeniable tracks in wet mud, exercising the imagination and keeping today's tamer West connected to a tooth-and-nail heritage.

The following morning dawns cloudy, the wind strong from the south and very damp, and the light is faint as we walk the trail to the tip of an anvil-shaped butte that juts out into the vee formed by the junction of Buffalo Creek and the Cimarron.

I'd wanted to see it at daybreak, because this spot is, according to a book about the Pawnee Indians, the place where ceremonies were held prior to the European invasion. Allegedly an ancient cedar tree grew at the base of this bluff, near where a salt spring gushed out of the vertical wall of red clay. We've searched for the cedar and the salt spring but they too have reverted to legend. Even so, this is still a powerful place with an enchanting view, even in the half light of a morning that promises an afternoon thunderstorm.

The throaty, liquid cooing of mourning doves carries down the ridgeline, and within minutes several birds land on a white rock outcropping overlooking the river. There's an old mesquite grove a hundred yards back from the rim of the bluffs and from within in it, a meadowlark serenade. These trees are stout with spreading limbs and long white thorns, widely spaced amid a meadow of new green grass. Many of the mesquite are dead or dying. Others have leaf buds soon to unfurl.

Diaries of cattle drovers indicate that trails up from Texas were grazed down so closely for a mile on either side of the main route that "not even a sheep could find a meal." It seems this was not an unusual occurrence, as cavalry troopers on patrol reported that herds of buffalo grazed the prairies so thoroughly that their horses could find little to eat in the great herds' wake.

Obviously the grasslands were accustomed to a periodic close shearing, from buffalo, from prairie dogs, and in the end, from vast herds of Indian ponies. Before fences, the animals moved on and the land recovered. Cattle changed all that as grazing pressure became constant and the root systems of native plants weakened under stress.

As a consequence woody species invaded, including mesquite. In modern times the spread of eastern red cedar, fanning out from native habitat along the rugged river breaks, threatens to overwhelm pastureland.

Overgrazing is part of the problem, as is absence of fire. The prairies evolved under a burning regimen, either from fires set by lightning or by Indians hoping to concentrate wildlife. Today much of the range in this area is in need of prescribed burning to control another invasion of woody plants.

As the day warms several turkey vultures ride thermals rising off the river valley, and their drifting flight reminds me of the way bonfire ash ascends and seems to hang above a hot blaze or even a kite without a kite string, effortlessly adrift. An afternoon of photography flies by quickly, and the day's signature is a fitting one, a turkey hen caught in sunset silhouette atop a roost tree, then a jackrabbit seemingly fleet as an antelope as it races in the headlights ahead of the truck. At dusk we stop to open a gate beside some old cattle working pens in the river valley and are overwhelmed by nature's finest perfume emanating from flowering plums, natives that have reclaimed the weathered planks and turned them into a work of art.

Ten-fifteen in the evening, and a warm golden glow creeps over the Buffalo Creek valley. The full moon climbs, soft light floods the cottonwood limbs, a few thin clouds drift across the sky for emphasis. Now all we need is an encore from the orchestra. Somewhere in the distance coyotes rosin up their bows, right on cue.

By mid-April, spring here in the Selman pastures seems less tentative, although the wind still pounds as it did in March and we're assured by precedent that these warm winds can suddenly turn and sweep down from the north, even usher in a belated snowstorm.

But for now the narrow box canyons hidden amid the Cimarron River bluffs are a jungle of color, with tangles of flowering currant, plum, and threading grapevines climbing all the way to the top of the steep slopes.

The gypsum caprock contains all sorts of eroded niches and mini-caves, and a porcupine, seemingly as big as an Airedale, rests on its haunches in one of these shelters, watching us struggle up a deer trail through the ensnarling grapes. The animal's morning lair is just above the spot where a spring erupts from one of the bluffs and cascades downhill between a pair of cottonwoods that have managed to take root on the nearly vertical embankment.

Up on the high prairie a northern harrier sweeps past riding the stout south wind, low to the ground and somehow more like a bat in flight than a bird. Moths are erupting from cracks in the chalky cliffs behind us, and new spring flowers are underfoot: a tiny parsley with carrot-like leaves, a butter yellow flower as simply shaped as a child could draw it, the pale lemon yellow of Indian paintbrush, a spreading legume with purple pea-like flowers, one of the locoweeds.

Later in the afternoon we drive north across the Cimarron to another large ranch, seeking a lonely prairie cemetery marking the spot where two cowboys were killed by Indians. We find the grave on a big, rolling plain sprinkled with a ground-hugging pale blue flower featuring multilayers of petals and sepals. It's one of the morning glory tribe with the less than lovely name of Evolvulvus. In this case the blooms brighten a lonely gravesite retaining both the original tombstone, epitaph carved in rock, and a modern stone that serves as a historical marker.

The original stone is the most poignant. A rough engraving simply says, "Cowboy And Salt Haul Killed by India."

The time was 1874, and the cowboys were driving a wagon south for a load of salt from the Cimarron. What they didn't know was that Dull Knife's band of northern Cheyennes, incarcerated with their southern cousins at the Darlington Agency near present-day El Reno, Oklahoma, had left in the night and were headed home.

Dull Knife's band was dying of disease, starvation, and maybe even a bad case of acute homesickness. The northern bands were distinct from the southern Cheyennes who hunted along the Arkansas River and south into the country along the Cimarron and Canadian. They visited back and forth, but Dull Knife was at home on the northern plains of Montana. The heat and the brackish water here in the south was no substitute.

Graft by Indian agents made it hard for the Cheyennes to procure even the most basic necessities, and game was scarce to nonexistent on the reservations. So Dull Knife's band hid weapons in the sandhills across from the agency and broke out after darkness, intent on following the rivers they knew back to a familiar homeland.

Unfortunately for the two salt haulers, the Cheyennes were desperate and left no witnesses to spread the alarm. The runaway band fought with military patrols throughout their exodus back to Montana and absorbed heavy losses. The Cheyennes were eventually captured once again. However, this time their flight, and the indignity of their plight, aroused a certain amount of public sentiment and eventually the northern tribesmen were granted a reservation of their own within their traditional country.

Hunted down like livestock that had slipped some bureaucratic traces, Dull Knife's running battle with those who sought to enslave him made a hero of the determined chief. Some of the West's best writers chronicled the event, and eventually the great John Ford would address this sorry episode of American history in a classic movie.

By late afternoon the wind is literally howling, with low gray clouds racing overhead as we drive down to a secluded spot along the banks of Buffalo Creek to view the remains of J.O.'s old home site.

The dilapidated frame house is surrounded by a grove of Siberian elms planted by J.O. when he first decided to establish a home here. The trees are now closing in on the century mark and show the signs of their aging. Broken limbs litter the ground and several appear ready to come crashing down during the next strong thunderstorm.

We find the rusted remains of an old wind generator, and Sue Selman mentions that J.O. used wind power to feed a crude storage battery that supplied electricity—a sign that J.O. Selman wasn't afraid to confront the possibilities of technology, a presence of mind that certainly kept him well ahead of his time. Today the doors are standing open to the old home place, and it seems quite large, considering the small family. Cattle have come inside and sought shelter from harsh weather, and that also seems fitting, as is the presence of a phoebe inside the big screened-in porch, along with wrens in the kitchen and barn swallows darting from under the eaves.

The railroad right of way came almost through the backyard, and J.O.'s shipping pens are only a few hundred yards away. All in all it seems the ghost of a well-ordered place, one that must have been immensely comfortable for the times, nestled in a beautiful setting along the cool banks of Buffalo Creek. I sense that "Little Jimmy Few Clothes" felt at home here, and I can only imagine the domestic battle that arose when Lena decided that with the children nearing school age, it was time to move to town.

The hills around the old Selman Ranch are big and rolling, divided by steep, deep draws cut by little spring-fed creeks bordered by cottonwoods. It seems a good place to watch closely for wildlife, and we're rewarded when lesser prairie chickens explode from some low, bushy cover and rocket toward a distant knoll like overgrown quail.

Prairie chickens are more accurately prairie grouse, and the lesser chicken is at home in the sand sagebrush, shinnery oak, and midgrass prairie country of western Oklahoma, southwest Kansas, southeast Colorado, the Texas Panhandle, and eastern New Mexico. A close cousin, the greater prairie chicken, ranges from northeast Oklahoma north through the Flint Hills of Kansas into Nebraska and the Dakotas, wherever prime prairie habitat remains. Another similar grouse of the Texas coast, the Attwater's prairie chicken, clings precariously to remaining habitat near Aransas Refuge. The heath hen of north-east coastal islands, another cousin, is already extinct.

Prairie chickens have exacting habitat requirements, namely countless square miles of native prairie grasslands with a certain amount of both grassy and weedy cover intact. Lesser prairie chickens have adapted to a more "brushy" prairie that can include low, dense mottes of shin oak and also the sagebrush hills frequently found in sandy country.

Unfortunately for the chickens, much of their range has been chopped up into domestic grain fields, especially where irrigation pumps can gain access to the vast Oologah Aquifer, a huge underground ocean of fresh water underlying a large chunk of the southern plains. With native prairie isolated into much smaller parcels and often overgrazed, the chickens have had a hard time of it and edged precariously close to the endangered species list. Once hunted extensively, lesser chickens are now protected and landowners like the Selmans are cooperating with U.S. Fish and Wildlife biologists in an effort to keep the animal off the endangered rolls. Several have entered into a gov-ernment cost-share program that provides help with cross fencing in some of the big-ger pastures and other projects beneficial to long-term range management.

By breaking down the size of the pastures, cattlemen can practice a grazing regimen that more naturally duplicates the buffalo grazing pressure these grasslands evolved under. Steers are turned out on a targeted pasture to graze extensively for a short period, then removed to the next pasture, and the next. In the meantime the closely cropped grass is left alone to recover until the following year.

This is a much more natural cycle, and it provides better native forage health and utilization by grazers. The system also allows for additional nesting and escape cover for species like prairie chickens and bobwhite quail. Biologists hope to enroll and enhance as many acres as possible, and eventually stabilize lesser prairie chicken numbers. The endangered species program has never been popular among landowners, since it carries with it federal mandates. Westerners, especially ranchers, tend to be personally and politically independent and regard government, even without mandates, with suspicion. Fear of environmental regulations in this region borders on psychosis.

So after years of conflict, federal agencies are beginning to understand that volunteer, cooperative ventures to save wildlife species are infinitely more successful, especially those that provide money for habitat improvements that the ranchers more than likely couldn't manage to underwrite on their own. In the end, both the biologists and the landowners benefit from such common-sense approaches . . . and vanishing grassland species find themselves the beneficiary of a little more time on this earth.

Today the chickens near the old Selman Ranch catch a strong south tailwind and glide north, toward a lonely little knoll in the middle of a pasture that appears to have the chop and roll of a restless ocean. New gray-green leaves adorn the sagebrush, and it is certain that these chickens meet on some nearby lek, or booming ground, early in the morning to participate in the strange courtship dance peculiar to the prairie grouse family. Males indulge in a wild display of cackling, cooing, foot stomping, wingtip dragging, and sometimes, airborne confrontation. Females look on calmly, eventually choosing a mate from among the gyrating suitors. It is a dance as old as the prairies, and once the chickens descended in huge flocks to perform it. Today on the Selman place you'll find several leks, yet most attract only several dozen birds at best. Prairie chickens are socially affixed to leks, and genetically linked birds will use the same dance arena down through generations. Modern booming ground censuses are discouraging, and it seems that for a few years at least, while available habitat improves, low numbers will be the norm and those who love these birds will pray that catastrophic storms or drought won't reduce the population even more.

The lesser prairie chicken isn't the only rare animal here in the midgrass region. The swift fox, long-tailed weasel, meadow jumping mouse, and black-tailed prairie dog exist today at only a fraction of their former numbers. Rarely seen birds include golden eagles, western burrowing owls, ferruginous hawks, mountain plovers, and long-billed curlews.

The list extends to critters rarely observed and less often recognized, like the Texas horned lizard, northern earless lizard, and the western Massasauga rattlesnake. Some species, like the black-footed ferret, are gone from southern plains grasslands due to the decline in prairie dog towns. Ferrets prey on prairie dogs, a species routinely eliminated by ranchers concerned by the amount of forage consumed by these social rodents that once existed in towns extending for hundreds of square miles.

Prairie dogs and their intricate system of underground burrows provide food and preferred habitat for a number of grassland species like burrowing owls. Ferruginous hawks, golden eagles, badgers, and rattlesnakes prey on dog town residents, and birds including mountain plovers and long-billed curlews are attracted to closely shorn townsites.

Reptiles like Texas horned lizards are at home on disturbed, sandy areas utilized by prairie dogs, especially where the mounds of harvester ant colonies are plentiful to supply these prickly critters with their favorite fare. Today some ecologist believe that the short, tender successional grasses that sprang up in the wake of a thorough dog town cropping actually drew buffalo to graze the perimeters of these once vast colonies. However, you'll rarely convince a rancher of this, because cattlemen look at a dog town and see grazing land so closely manicured that it resembles a golf green.

Certainly prairie dogs played a key role in creating an ecosystem so diverse and rich in animal life that early explorers would refer to the plains region as America's Serengetti. A glance at historical records just on the Selman Ranch alone indicates the region was once inhabited or visited by buffalo, elk, mule and whitetail deer, black bear, maybe an occasional grizzly, bobwhite quail, turkeys, lesser prairie chickens, prairie dogs, and with them more than likely swift fox, burrowing owls, mountain plovers, and curlews. This winter both golden eagles and prairie falcons were seen on the ranch, but one can only guess at how much the actual numbers have diminished.

The problem with preserving prairie wildlife is perplexing due to the scale required to reestablish a functioning ecosystem. Simply put, prairie species evolved on vast grasslands that stretched unbroken from the Crosstimbers region dividing Texas and Oklahoma west to the foothills of the Rockies and north through Kansas and Colorado into Nebraska, the Dakotas, and parts of Wyoming, Montana, and Canada.

Within this unimaginable space, prairie dog towns could exist within a hundred linear miles and still have plenty of room to migrate to some other location and start anew. Fires started by lightning or Indians might burn untold thousands of acres, yet millions of acres still remained untouched. Drought could parch a region as big as the Texas Panhandle while a summer deluge could paint what is now western Oklahoma emerald green.

Essentially the plains were like a vast quilt. Though one unified whole, the surface area was patchy due to localized conditions. Plant life occurred in a variety of succession stages, brushy here, cropped by prairie dogs there, burned by Indians in places, grazed by buffalo in others. In some areas forbs dominated, in others grasses. Herds of grazers were constantly on the move, searching for and utilizing species that best served their seasonal nutritional requirements. It would have been akin to visiting the African grasslands in their prime to have seen all the deer, elk, antelope, and buffalo that roamed America's plains, along with the coyotes, wolves, and bears that followed.

Today these same plains are chopped up into ranches interspersed with farmland wherever water is available to grow crops. Cattle operations like the Selman Ranch may seem big to a visitor from New York, but 16,000 acres remains only a speck of the original prairie that stretched over half a continent and did not know a fence.

We can't tear down the fences and replant all the wheat and sorghum fields to native grass. Therefore, it remains impossible to restore prairie wildlife to naturally renewing numbers. As it is, species like prairie chickens increasingly cling to islands of native grassland amid blocks of farmland. This hampers the genetic mixing required for good species health and makes it difficult for the birds to weather bad years. In the past, the flocks might have simply moved to another area and then recolonized when conditions were favorable. Now the chickens are locked in place and susceptible to drought, fire, and overgrazing. As biologists found with the heath hen, a quick series of unsuccessful nesting seasons can cause isolated chicken populations to spiral below the point of recovery. Places like the Selman Ranch offer the last best hope, but even then hundreds of thousands of additional acres within reasonably close proximity are needed to keep populations viable on a regional basis.

May on the plains can be a spring month or a summer one. Cool, gray days keep grass growth in check, so ranchers prefer wet and warm, with emphasis on wet.

By May 14 we find that the winds are still blowing at speeds up to 30 miles per hour out of the southwest, except now it's 90 degrees. Big, puffy white clouds sail overhead, casting shadows like ships on the green ocean of new grass. The Selman place received four inches of rain just one week ago, but the fierce gales have already sucked the moisture from the surface and dust dances across the yard as Sue Selman's pack of dogs chase first a Frisbee and then one another.

Now is the time to notice new calves in the pasture because, quite simply put, few animals are more beautiful. Calves here on the ranch are a rainbow mix, featuring various combinations of black and white, black with white faces, and the traditional brindle-brown calves, the latter easy for my love of longhorn color patterns to accept as the best of the lot, even though these calves are mostly of Hereford and Angus blood . . . unless some longhorn bull jumped a fence.

In a way there's a touch of antiquity to the scene we witness along Sleeping Bear Creek. Bulls, cows, and calves are grazing contentedly while swallows swoop low to ingest the flies that plague each animal. This is symbiosis at its best, much like cooperation between neighboring ranchers when a need for manpower arises. By sharing time, energy, and skill, several ranch families can manage thousands of acres at a profit, whereas a payroll outlay for the equivalent in hired help would stretch the budget and in hard times, snap it.

Ranchers revel in their independent image, but from the first they've been a communal lot, given to a particular mode of dress and cultural habits as narrowly defined as that of any band of Comanches. Many tend to think that the cowboy image mostly has its roots in cinema, but those who've grown up around the western culture realize that the style itself is almost sacred and evolved long before Tom Mix or John Wayne. Every item of clothing, every tool, evolved out of practical necessity, from boots to spurs to long-sleeved shirts to broad-brimmed hats. The only thing that literally doesn't fit are the tight blue jeans. The old cowboys I grew up around wore loose-fitting khakis, and if you'll notice the cowboys pictured in historical photos in this book, loose-fitting khaki-type pants were the norm. Jeans are miserable when wet, and today's tight-fitting Wranglers make it almost impossible to swing a leg over a saddle. However, tight pants look good on the young rodeo stars and women are prone to notice such things, as are marketers and advertisers.

Ranchers' kids I knew growing up were almost always very worldly and capable at an early age, as well as hard working, dependable, soft spoken, polite, and finally ornery as hell on the weekends when the beer flowed and inhibitions departed. More than anything else, the kids were tough, physically and mentally, and that goes for either gender. They dressed the part of the cowboy culture, lived it, believed in it, dedicated their lives to it. Unfortunately, economics would force the majority away from a lifelong commitment to cattle and horses.

The scarcity of ranch-related jobs and the fact that young wives are no longer content to live on a few hundred dollars a month and all the beef you can eat have delivered a number of young wannabe cowboys into trade school and blue-collar jobs they hope they can somehow tolerate. However, come Saturday night, the progeny of the prairie break out the costume and are once again of the same spirit as the horse and the prairie winds . . . even if that wind blows from a ceiling fan in some Oklahoma City saloon.

Now that the Selman Ranch has received its spring blessing in the form of the recent thunderstorm, all the creeks are full and running strong, just deep enough for the Selman clan to float downstream in truck innertubes on their annual spring catfishing expedition.

Buffalo Creek produces fat channel catfish, and they bite particularly well on salamanders the Selman boys seine from a neighbor's pond. At the end of the day the catfish fillets go on the grill, cold beer takes the sting out of the inevitable sunburn, and friends and family celebrate ranch-style. City kids may sniff at such festivities, but the junior members of this clan have spent more time living in cities than here on the ranch, and they obviously prefer such simple pleasures. Everyone has heard clichés about how the land gets in one's blood, but today we see evidence of it. The younger Selmans may have broken ranks with the traditional style of country living, because you see more tee shirts and running shoes than boots and broad-brimmed Stetsons. Nevertheless, you also sense that these young people belong here. The old styles may catch up with them at a later date. But for now it's good to watch a new generation connect with the land. Without that connection, this marvelous patch of prairie has little hope of surviving in essentially the same shape that J.O. Selman found it in more than a century ago.

Bob Selman loved to operate heavy machinery and build ponds, and the ones on this ranch are beauties thanks to his careful work and the numerous springs that occur in the Buffalo, Sleeping Bear, and Sand Creek drainages.

The ponds are deep, clear, great for swimming, for fishing, for waterfowl resting.

This April morning several of the ponds hold bachelor groups of mallard drakes, most likely young males that failed to secure a mate and may spend the spring here in the South, simply loafing.

Killdeer frequent the barren banks of these impoundments and scream their displeasure at our intrusion. Rather than hassle them further we decide to take in the view of the Cimarron salt plain from the big bluff that divides the major river valley from that of Buffalo Creek. Strong thermals arise from the Cimarron and provide a favorable climate for soaring turkey vultures. They drift in circles, ever higher, seemingly into the clouds and out of sight. Our daydreams tend to follow.

Greenup is well underway and in places the grass is lush. In others grazing pressure has left little other than a thick blanket of woolly plantain. And of course we find plenty of the introduced brome called cheat, plague of the prairies. There's also a splash of color from a very lovely low-growing wildflower, red tipped with yellow. The common name is Indian blanket and it fits. Nearby we discover two more species of gaillardia, one totally lacking in ray flowers and thus the bloom's showy floral display, and another known as cutleaf gaillardia, with fewer bright ray flowers around the plain brown disk flowers, those of little sex appeal and all the fertility.

Just at sunset we stop to photograph a thick colony of soapweed yucca. The tall flower stalks now bear big, cream-yellow, almost lascivious bell-shaped flowers in heavy clusters. This colony, in a portion of highway right of way that escapes the mower, is perfect for the camera. Those growing in the pasture have been grazed or browsed and oftentimes a diminished flower cluster emerges on half a stalk or stem, below the amputation.

We have enough light left to examine the slim flower clusters, each several inches long, now appearing on the scattered mesquite trees. The flowers appear with the new leaves and, in this late evening light, the entire effect is as if the trees and their flowers are dripping with liquid gold. Less obvious yet certainly more common are the antelope horn milkweeds with inconspicuous yellow flowers. The yuccas, much more brazen with their display, depend upon a certain species of moth to achieve pollination. The less showy milkweed flowers must spin a narcotic spell with their fragrance, for they seem overwhelmed with swarms of small brown skipper butterflies.

May hopefully is a month of rain and most always is a month of wildflowers, and this year the Selman Ranch basks in the serenity of both.

A ride across the high, rolling hills overlooking Buffalo Creek near its junction with Sleeping Bear on this mid-May afternoon would stir the hearts of any wildflower lover, and I imagine it is a spectacle that provides good medicine for the hearts of crusty old ranchers as well, though they would be loath to admit it.

It is difficult to ride past a cluster of big penstemons without stopping to admire the shameless seduction of the bloom. This white flower is big, maybe an inch across, bell shaped with the throat curled back just enough to display dark purple streaks leading deep into the blossom, like landing lights on an airstrip, guiding insect pollinators to the sweet secrets within.

Some of the penstemons are darker, almost purple, and while they are simply pretty to our eyes the color scheme has more to do with ultraviolet displays that catch the eye of insects. The breaks along Buffalo Creek tend to outcrop in the form of little knolls creased by deep ravines, and the penstemons cling to the eroded red dirt banks of the steepest slopes. In the late evening light the scene takes on magic, and it's easy to imagine a Cheyenne medicine man atop one of the knolls, sitting there alone with a cedar flute or small rawhide drum. From such a vantage point one can see for miles, including the silver ribbon of Buffalo Creek, patches of red dirt, new green grass, white gypsum outcroppings, and these lavish purple and white flowers. A clump of penstemons on a particularly steep red dirt bank has summoned a big yellow sphinx moth that hovers around the flowers. This insect darts from bloom to bloom like a hummingbird, and in fact is oftentimes confused with the same, thus another colloquial name, hummingbird moth, in this instance a name that fits.

Our ride reveals a veritable carpet of wildflowers, mostly low-growing white daisies and an assortment of legumes including a reddish purple locoweed. However, the season of taller flowers is at hand, and we find a Baptisia, known locally as wild blue indigo, with a purple pea-like flower that produces a black seedbox that rattles in the wind.

The Baptisia grows knee high, along with a white larkspur, the flower named for the way the flower tapers to a "spur" like that of a bird's foot. Scattered amid the ground-hugging flowers is a taller, butter yellow daisy, bright neon yellow flax, and the pale lemon tint of yellow paintbrush, so thick in places that entire meadows nestled between the stark knolls appear to have been spray painted this lemony hue.

Before heading back to the ranch we take time to visit the junction of Sand and Buffalo Creeks, where an ancient elm clings to life by the thread of a taproot after being downed by a fierce windstorm. Only a sliver of wood several inches across remains imbedded in the earth, yet the huge tree is in the process of manufacturing new leaves. We admire the effort, but wonder if the hot winds of summer won't end the struggle once and for all.

Beavers have built a small dam on Sand Creek in the shade of a soapberry grove. The banks bordering the beaver pond display rich dark soils, sprays of hot pink sensitive briar, and the dark purple blooms of spiderwort. A spotted sandpiper, black spots against a white breast that bobs like an oil field pumping jack, appears alarmed by our presence yet flies only 30 yards or so up the creek before landing to indulge in more nervous dipping. These prairies produce prodigious numbers of insects and the insectivorous birds are active now, gleaning protein on the wing.

Nighthawks have vacated their midday fencepost perches and circle overhead, sipping supper out of the damp evening air. Earlier we saw a Mississippi kite on the wing near the Selman's south fenceline, a bird of aerial insect-catching acrobatics that rival those of the nighthawks. And, in the ancient Siberian elms at the Selman place, great crested flycatchers squawk at the world around them from the upper limbs, the voice raucous, the ability to dart from a perch and snag an insect adroit.

In a barren patch of sand near Buffalo Creek a Texas horned lizard scrambles to acquire its daily ration of nearly a hundred harvester ants, while a few feet away a pair of dung beetles struggle to roll a tidy little ball of manure to a place of safekeeping. The manure will nourish the beetles' offspring when they hatch, but first the larder must be moved from the area of disposition to the site of safe haven. So the two insects, neither bigger than my thumbnail, struggle to push their prize up a steep incline maybe a foot high. One beetle shoves with its back legs while the other pulls with its front. The dung ball slips only inches from the top, and the insects repeat the process, adding emphasis to the meaning of tenacity.

Just at sunset we watch a turkey hen ease through the sagebrush and enter a soapberry grove along a little spring-fed rill feeding into Buffalo Creek. These western Rio Grande birds are leggy and remind me a bit of ostriches. In this case the turkey's careful retreat, maybe to a nest, is a fitting close to a perfect May day. This is the time when ranchers sit on the back porch at sunset and thank God for the real currency of the plains, sunlight, rain, and the green grass that's bound to follow. Once again the coyotes voice their approval, and we retire to the bunkhouse with appetites earned through honest outdoor toil.

≈≈≈≈≈≈≈≈≈≈≈≈

From a distance the pastures on the north side of the Selman place appear to be mostly high and rolling, devoid of contour, offering little in the way of the spectacular scenery found along Sleeping Bear Creek.

However, a half-mile hike reveals several narrow canyons, each lined with cottonwoods, all cut by perennial springs. Gypsum outcrops overlook the headwaters of each spring, and in places we find chunks and flakes of chert, a beautiful stone of a rich purple hue that practically looks hand polished.

It's easy to imagine a native hunter stopping here, sitting with his back propped against the white rock ledge in the shade of one of these massive cottonwoods, cold water trickling out of the stone maybe ten feet from his moccasins, pressure applied from a deer antler popping off thin slivers of chert from the piece he has selected to make a quick cutting edge for butchering a fat doe that lies dead near the plum thicket above the spring.

We would like to find projectile points, and someday I feel we will, because this country has that gamey feel to it that hunters understand and savor. A grapevine tangle overhead holds the promise of sweet fruit, and in the dark mud along the coldwater rivulet, deer tracks abound.

≈≈≈≈≈≈≈≈≈≈≈≈

This May morning dawns with scissortail flycatchers fighting a stiff wind to snatch insects from just above the tussocks of prairie grass, while eastern kingbirds, another of the flycatcher clan, take time out to harass a hawk that has snagged a mouse from the roadside right of way.

We've gone to the eastern boundary of the ranch to photograph a windmill, awkwardly constructed yet still churning out fresh water into a rusted trough, the haphazard way the boards have been nailed together indicative of how much cowboys disdain work other than the hours spent in the saddle.

A neat row of mulberry trees on a gentle incline above the windmill suggests that this was a pioneer homesite, though little remains other than the trees and their ripe fruit. From here the land slips downhill several miles to the salt flats that creep up the mouth of Buffalo Creek. Between the mulberries and the silver sheen of the stream we see the rusty remnants of an old threshing machine, the iron wheels locked in place by time. Plant residue left in the hopper has turned to rich earth and supports a vigorous prickly pear cactus. It is one of the most demonstrative decorative planters I've ever come across.

The plains above Buffalo Creek display the pink flowers of a native mimosa no more than three feet tall. The floral display of a little roadside cemetery is equally lavish, an old-fashioned prairie tribute to the souls that rest here.

Gravestones are handmade, either of native stone or concrete, as simple as the lives that came to rest. John and Lovina Daniel lie here, born in 1821 and 1829, gone in 1909 and 1910.

Mittie Lee Leflors rests here with John and Lovina. Mittie Lee, born in 1887, had her life cut short in 1914. We pay our respects to Edith Moore, laid to rest in 1918. And Nathan Watkins, 1842–1907, "at rest" inscribed on his crude concrete marker. And Arthur Watkins, born 1906, gone from this world in 1907. The sense of tragedy is almost as unbearable as the wildflowers are beautiful and the headstones crude.

So we observe a moment of silence in honor of the sheer courage and strength that emanates from the souls buried in this cemetery. Soon it will be Memorial Day, but this year decorations are already in place: spiderworts the rich purple of princely robes, the nearly translucent yellow of paintbrush, larkspurs like snow at the end of a knee-high stalk, the neon butterstain of false dandelions. Clumps of buffalo grass hoist little flaglike flowers on stalks from six to eight inches tall, and coyote gourds wrap around the weather-stained stones while tiny wine cup mallows underscore the inscriptions. Spring Valley Cemetery, the sign says, here on a windswept hillside with no sign of a valley or a spring. However, we understand now that the land reveals its secrets to those who go in search of them, and we are certain that beyond the horizon awaits a sheltered valley with a sweetwater spring, and that these people named on these stones left something of themselves in that special place, as each passed by into eternity.

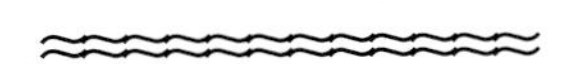

June is a lush month on the Selman Ranch, a time of phenomenal growth if spring rains are kind. Even this far west June can be both hot and humid, before the dry southwest winds of July couple with 100-degree temperatures to turn the midday hours into an open-air furnace.

June is the month when spring calves fatten, when fruits ripen, when the land is most fecund. You can almost feel the life swelling all around, the earth trembling with effort, the insects hatching, the nestlings fledging, the native grasses stretching out to touch the sun as the rancher tinkers with his baler and casts an eye over the hay meadow, mentally calculating the tonnage he'll store.

It is now June 20, the time of the solstice, nearing the moment when daylight will reign over the ranch for the greatest number of hours, a holy day in times past when so-called primitive cultures were more attuned to the movement of the planets and the effect these celestial meanderings had upon their daily lives.

Of course the summer solstice still impacts each of us in profound fashion, yet now our teeming cities bulge with billions connected to a life support system created by technology, totally out of tune with the ancient rhythms that account for bacon, eggs, bread, and of course burgers, the end result of the grass that began to grow here at the spring equinox and fattened the calf that will travel from native prairie to feedlot for grain fattening before ending up in a paper sack thrust out a drive-in window to a harassed housewife during rush hour.

Last night our rendezvous with primordial planetary forces was awarded the added emphasis of Mars reigning dominant on the southeastern horizon. No mistaking this planet for something more mundane, like satellite or airplane. It seemed to stand there, arms folded across a massive red chest, sizing us up, glowing brighter than a bomb blast, red tinted like the red dirt of the bluffs just beyond the county road.

It was enough to keep us talking late into the evening, and the lights in the bunkhouse buzzed with all manner of bugs. Sue Selman's grandfather Cowan died of emphysema in this place that now serves as guest house and hunting lodge. Sue says his fingers were brown as shoe polish from the nicotine, and that he smoked hand-rolled cigarettes right up until the end. Tonight I believe he is ready for us to retire to bed so he can sit on the porch for a while and savor the spectacle of moon and stars, and of course Mars burning brightly over Buffalo Creek. Sue tells us his spirit remains fond of porch sitting and savoring the night sky.

A light shower overnight cools the air enough to keep the rattlesnakes from roaming far. Early the next morning we drive the loop road east to a point where Buffalo Creek and the Cimarron merge, stopping to visit the old schoolhouse where youngsters from Salt Springs congregated more than a hundred years ago.

Photographer Don House is a man in love with antiquity, in the way it reveals itself in sagging structures and weathered boards. We leave him and his camera alone for reflection and introspection, for he understands that cameras readily capture ghosts if each image is carefully composed and the time of exposure properly chosen. He does not want or need company as he divines for the spirit of these children and the dedicated woman who gave them not only multiplication tables, but also dreams that reached beyond their next sparse meal.

I doubt that any real cattleman can resist riding pasture when the native grasses are green and robust, and a lush growth of prairie forage covers the earth here between these tight fences. The pasture surrounding the old school shows all the positives of careful management, and we walk through shaggy clumps of buffalo grass, sideoats and hairy grama, all hoisting tiny brown seed flags.

Gourd vines have spread and bear either big, yellow squash-like flowers or golf-ball-sized fruits, now dark green with pale green stripes, soon to ripen with a yellow tint, big as apples. Bush morning glory flowers are stunningly purple, and the sandhills are yellow with tall primrose.

You can't consider a prairie June without dwelling on plants, particularly the tall wildflowers. Even the most hardbitten old cowboys have a soft spot for this time. Rather than wax poetic they just saddle up a horse, call the dog, and go for a ride, or drive the pickup very slowly down the pasture roads. These are solitary people in a solitary landscape, and what they feel and what they see are often one and the same, with little need to elaborate on it.

Late June is a time of legumes, the wild peas and lespedezas, the whites and lavenders of the wild alfalfas, the vital seed crop that Selman quail will turn to in the autumn. And with the flowers come butterflies, swarms of them, yellow and white like the dominant flowers. At our feet we notice something rustling in the grass and a hen lark sparrow attempts to feign injury and lure us away from four eggs, creamy white with black irregular spots like paint splatters, all tucked tightly into a tussock of bluestem.

A June prairie can be so saturated with life that it almost overloads the senses. A pale gray Mississippi kite sails overhead, and bright yellow prairie coneflowers are underfoot. A dicksissel, yellow breast reminding me of a miniature meadowlark, sings lustily from a low bush. A sharply whistled blast from a bobwhite quail, carving the air so cleanly, a longhorn bull in a pasture of Hereford cows the color of the red clay on our boots, soapberries along Sleeping Bear Creek with yellowish green flower clusters, plums beginning to ripen, clusters of green grapes not yet ready for wild appetites.

The colors of late June in this "bland" landscape are yellow and green, purple and white. They include the bright yellow of sunflowers, cutleaf daisies, yellow woolly white, and a wiry little aster growing in tight clumps. For contrast the prairie offers the pale lavender of monarda or horsemint, the bright pinkish purple of the tall thistles, the deep purple of prairie clover, the rich red-purple of little mallows locals refer to as cowboy rose. White is common, as witnessed in the flowers on guara or the low-growing crotons presently being chewed by caterpillars with apparent gusto.

Late in the evening we watch the sundown ritual of both whitetail and mule deer does. For the deer the colors of various plants represent nourishment. For the yellow-billed cuckoos in the elms back at the ranch, these same colors represent nutrition once removed in the form of insects attracted to the plants. For the inch long, black and orange cerambycid beetle feasting in the folds of the huge white flower of prickly poppy, the colors signal sweet nectar. The beauty of prairie wildflowers provides a practical way for these plants to procreate, pass along genes across generations, no different, actually, than the biological urges of a beautiful young woman in her favorite party dress. Reproduction is the primary goal of beauty, along with many fringe benefits, especially for insect- and seed-eating animals. Nor does the circle stop here, for in turn the critters that eat insects and seeds are fulfilling a role in overall flower health both by controlling the bugs and by scattering seed. Everything here is connected in some way, even to that of the fashionable beauty who may find some of these herbs in her medicine chest, or the fragrance of one of these plants emanating faintly from behind her ear. And if some young man finds her favorable, the process begins anew, just as it does when the hummingbird moth dips down to draw sustenance from the penstemon.

This evening we visit the bat caves on Trader's Creek, part of the Green Selman Ranch. Green was J.O.'s brother and also a trail drover who, according to family legend, herded longhorns as far north as Montana.

The cave in the gypsum bluffs above Trader's Creek serve as a maternity ward for millions of Mexican free-tailed bats, a species found in the southern United States, Mexico, and Central and South America. Females migrate here in the spring, raise their young in the cave, and then drift back south to winter caves in Mexico.

Maternity caves are found in Texas, New Mexico, Arizona, and Oklahoma. The largest known maternity cave is near San Antonio, Texas, with peak summer populations estimated at some 20 million free-tails. Biologists estimate that the first 80 feet of the cave on the Selman Ranch hosts at least a million bats.

This may sound like a lot of bats, but as it stands, overall numbers are declining. Females tend to return in large numbers to a limited number of brood caves, all subject to vandalism. Pesticides also take a toll, according to bat biologists.

The bats' worst enemy may be the movies, which typically portray these animals as inherently evil. In fact the tiny flying mammals, only nine centimeters in length and weighing 15 grams, damage little other than the imagination and, at the Selman maternity cave alone, consume some ten tons of moths, mosquitoes, and beetles every evening, flying as many as 31 miles from the cave to snatch insects from the air.

The free-tailed bat life cycle begins anew in late March and early April, when mating takes place. Following mating, the females journey to the maternity caves where, after a gestation period of 11 to 12 weeks, all the pups are born during a two-week period in midsummer.

The pups take their first flight from the cave at around six weeks of age and may live as long as 12 years. When the flights peak in late July, the stream of bats from the Trader's Creek cave can seem surreal. The late evening sky over the cave becomes a twisting swarm of hungry free-tails so thick that at times they gyrate and darken the horizon like the smoke and ash from a huge prairie fire. Sometimes the animals form a fluttering pipeline leading to the rich foraging grounds of the Cimarron, several miles away. Others swoop low to lap up the rich bounty of mosquitoes hovering over Trader's Creek. It is not a good place for those open to the influence of sinister bat mythology, because the free-tails oftentimes dip down to suck up mosquitoes buzzing within ear range. Others more attuned to reality simply cheer them on.

On this particular June evening a few bats flutter from the cave just at dark. However, the major, massive flight doesn't emerge until after the sun has slipped behind the bluffs above Trader's Creek. In the background a chuck-will's-widow recites a song much like that of its eastern cousin, the whippoorwill. And, right on cue, great horned owls drop in to perch on nearby utility poles where they await this mass exodus that for the owls presents fast food on the wing. While we wait we're entertained by a horned lizard searching for ants in the parking lot. The lot is big enough for a bus because Betty Selman, public-spirited and conservation-minded owner of this ranch, had the vision to deed 340 acres around the cave to the Oklahoma Department of Wildlife Conservation after biologists determined the significance of the site.

Thus the maternity cave now receives permanent protection and, on weekends during the summer, visitors are escorted here by wildlife department personnel to learn about bats and witness the spectacle of their evening flight. According to old diaries, a buggy ride to the bat cave was a favorite excursion as many as 100 years ago. The fact that this cave has never been assaulted with explosives, as has been the plight of other maternity caves in the past, is a testimonial to the conservation ethic of the Selman family.

As late evening thunderstorms build over the bat cave and Trader's Creek trickles by clear and free, it's easy to envision the son of Daniel Boone camped here, seeking to sort out Indian legend from geological fact. Soon this stream and others coveted by J.O. Selman would become part of a family legacy that now stretches over three generations and the beginnings of a fourth. One can only hope that the spirit of this place will inspire the next tier of Selman heirs as strongly as it has guided the will of Betty Selman, a woman of vision. Due to her conservation ethic the biological integrity of this ranch will live beyond her, just as she will always be a part of this land.

The storm breaks a little after nine in the evening, and we're treated to an unforgettable scene as bats pour from the cave, illuminated by a barrage of lightning strikes. With each flash thousands more swirl heavenward, and we watch as owl wings mingle with bat wings and the silver sheen of the stream reflects a strobe-like series of flashes, followed by reverberating thunder that shakes the buttes that surround us. The rain comes down in torrents for several hours afterward, but by now we're safe in the bunkhouse, too enthralled to sleep, entranced by the light show that brightens the horizon in every direction. Tomorrow, if we sit quietly down by Buffalo Creek, I think it may just be possible to hear grass grow. A rain in late June is always heaven sent. The morning dawns cool, with a crescendo of birdsong.

Early August appears to be typical this year. High temperatures remain near 100, although not as hot as July. Winds blow typically from the south/southwest, around 10 to 15 miles per hour. The prairies are parched following a month of hot winds, high temperatures, little rain.

Every day at midday we watch the big cumulus clouds build in size and numbers, most cotton white, others tinged with gray. And we seek to manipulate them through sheer willpower to congeal into a single, dark, swollen mass that darkens the sun and drenches us with cooling rain.

Churches throughout the plains routinely pray for rain and, in drought years, the governor may even make such prayer an official proclamation. Southwestern tribes have ceremonies designed to beseech their gods to be merciful in the distribution of moisture, and while the folks in prairie country congregations may not have a special god just for rain or a ceremony that involves serpents, their prayers are as strong and tightly targeted as those of any primitive shaman.

However, for a while at least, rain simply is not going to happen. So we amuse ourselves with childhood rituals, including a search for familiar objects in the shapes of clouds, or watching the immense shadows sail across the grass ad infinitum.

After several weeks of extreme heat and drought the tufts of buffalo grass are beginning to go dormant. Grasses that evolved under an unforgiving sun and periodic dry cycles simply turn to their own form of hibernation and await the return of moisture. Other warm season grasses more acclimated to the heat may do little other than develop brown tips on the long late summer leaves. Much depends upon the grazing regimen and the amount of reserves stored in root systems.

Eventually severe hot weather and lack of moisture can desiccate even the hardiest range plants, and ranchers may be forced to supply cattle with supplemental feed as the pastures turn as dry as a bowl of breakfast cereal before the milk. Buffalo herds once had the freedom to migrate to greener pastures where the spotty summer rains had fallen, but that was before fences. The cruelest trick of nature may occur when a rancher peers out the window of his or her pickup truck and watches a thunderstorm produce a flash flood on a neighbor's property, while the home place remains tinder dry.

Even in the driest times August is not without beauty, for a number of late summer flowers draw upon stored reserves to bloom even when the temperature tops a hundred degrees, and the winds seem to suck moisture straight through the skin. Today the prairies are white with snow-on-the-mountain and golden from a healthy crop of sunflowers. These are all tall plants, waist high or more and growing up to six or seven feet.

In the sandy soil along the streams we discover another late summer wildflower, this one with a big, showy flower at least four inches in diameter. Stamens are several inches long and form a thick cluster that protrudes from the bloom's throat, giving the flower a bushy look. In late evening this plant, commonly known as sand lily, almost glows with a neon-like radiance.

Some of the grasses that grew so lavishly in late June are now heavy with seed. Sideoats grama has completed the annual cycle and seems appropriately named as the seeds look like oats and climb the length of the stem. Stems and leaves of bluestem indicate maturity with namesake powder-blue hues, and legumes like Illinois bundleflower, heavy with little "shaving brush" white flowers in spring, now carry clusters of small brown beans in namesake bundles. We'll find the bright orange-brown seeds in the crops of quail in November.

Despite the heat, all life forms appear intent on the process of replicating themselves. Harvester ants are busy storing food, and horned toads are equally busy crossing the powder dirt of ranch roads in search of ant trails. Dung beetles are still at work pushing tiny balls of cow manure to a secure resting place, and they work individually or in tandem. Tiny gnats fill the air near the mouth of Buffalo Creek, and I rub them from the watery corners of my eyes and try to focus on something other than irritation. Fortunately deer tracks are frequent here, and I'm able to imagine the huge bucks that will splash through the shallows in November.

In August the salt crust that wells up near the mouth of Buffalo Creek grows thick and extends over thousands of acres. At this time the stream cuts a swath through the salt, looking much like a little mountain creek rolling down through a field of snow.

We see yellowish, tiger-striped plains killifish and swarms of shiners in the clear, briny current. Nor are we the only searchers. Low overhead are dozens of least terns, hovering like helicopter pilots, agile on their long, slim wings. After a few seconds of flying in place a tern selects a target and dives to snare a minnow from the shallow stream. We see and hear the splash, then watch as the triumphant fisherman retreats, shiner clutched tightly in the long, slim beak.

Both adult and juvenile terns are loafing along the banks of Buffalo Creek, their foraging flights on hold. At the same time, numerous small sandpipers, those difficult-to-identify little shorebirds known collectively as "peeps," scurry along the edge of the stream in search of brine insects or tiny minnows. A few larger sandpipers feed downstream beside a snowy plover. Within minutes a greater yellowlegs passes overhead, rowing hard against the wind with tapered wings, lamenting about something in poignant, high-pitched shorebird fashion.

The salt is alive with tiger beetles, all seemingly on a mission, just as the grass growing at the edge of the salt pops and cracks with small yellow grasshoppers. We search the extensive brine pools for salt crystals and discover thick chunks that look like coral, while in other places the brine has evaporated in a lacy pattern. Oftentimes the crust appears to be solid, yet we find ourselves breaking through the surface into shallow pools that lie beneath.

Several miles upstream, where the salt content is moderated by an influx of fresh spring water, we watch several carp forage in a shallow pool, then a pair of channel catfish that might weigh in at several pounds apiece. All appear to be watched closely by schools of green sunfish.

The Cimarron seems a rather placid river until you experience a spring flood or become mired in the notorious quicksand. Legend has it that somewhere in the Cimarron rests a steam engine, swallowed up when a flash flood took out a railroad bridge. Other accounts tell of Indians driving willow poles into these quicksandy rivers of the plains, to mark a safe passage for villages on the move.

And, during summertime low-flow levels, the Cimarron oftentimes became so salty that drinking from the river was hazardous to the Texas trail herds. Bayless John Fletcher, a Texas cowboy who kept a diary while driving 2,500 head of Texas longhorns into the Cimarron country in 1879, remembers a particularly perilous river crossing.

The herd "reached the Cimarron River, which they crossed near the mouth of Kingfisher Creek," Fletcher wrote. "The water of this stream was low, barely flowing, and it was so salty that we scarcely could drink it. Soon after we had crossed to the other side, a horseman came to our camp and presented a letter from the owner of the cattle, a Mr. Snyder, authorizing the rider to pilot us through the region of the saline and gypsum waters. The rivers being low, the waters were strongly impregnated with salts.

"The pilot, Bud Armstrong, informed the group that at one crossing ahead of us on the Cimarron there was a saline deposit so strong that one outfit lost more than a hundred cattle which died from drinking the briny waters. The Chisholm Trail had avoided that crossing, but now southern Kansas, due north of us, had been settled so thickly with homesteaders that we would be forced to abandon the old trail and detour to the west. We would have to recross to the south side of the Cimarron at the saline reservation (salt plain of the Cimarron), where the waters were the worst. When we reached that ford, we must stampede the herd and run the cattle across the river without permitting them to drink. After nearly four days of trailing on the north side of the Cimarron, through a poor sandy region wooded with a straggling forest of oaks, we came to the saline reservation, where we were to recross the river without permitting a single animal to drink from the briny waters.

"Armstrong now had us round up our cattle into a compact herd on the north side. We must then get them into a swift stampede before they reached the water's brink. This feat was accomplished by whooping at them as we fought them with our slickers and whipping them with our ropes until they were fully stampeded. Now we turned them directly across the stream. Then we had to frighten them on. Since the water at that time was shallow and the riverbed was sandy, we had no great trouble in crossing at full speed. But once we were across our troubles began.

"The thirsty animals tried to turn back, and it was night before we got them a mile from the river. They were nearly famished. All night long they tried to escape to the river to drink. We had little rest that night. The stench from the putrid bodies of more than a hundred cattle warned us of the danger in allowing our cows to drink from the deadly waters.

"Keeping at a safe distance from the deadly stream with its wide channel resembling a field of snow from the salt crystals upon the dry sand, we came on the following day to Buffalo Creek, a fresh water tributary of the Cimarron. There we found abundant fresh water for our famished cattle and horses. Since pasturage was good, we decided to rest a day to recruit our animals."

So the Cimarron, which roughly translates to "wild" or "unruly" in Spanish, can be a killer. Legends have it that the sands have sucked down wagons and even a field howitzer. However, on this particular day the Cimarron country has been blessed by increasing cloud cover, enough to break the tepid heat. By midmorning a few sprinkles dimple pools in the tributary creeks, and in the distance storm clouds form towering blue mounds trailing paler pillars of rain. Much-needed moisture appears to reach the ground maybe 15 or 20 miles away.

Somewhere beneath those pale blue pillars, someone is celebrating. Here on Buffalo Creek, stalks of wild buckwheat bend in the wind and the lower limbs of the cottonwoods are lined with lark sparrows, active with the cooler temperatures. For a few hours at least wildlife activity increases: a thirteen-lined ground squirrel dashes across a ranch road in front of our pickup truck, cottontail rabbits dart for cover in the sagebrush, a jackrabbit races ahead of us on the two-track trail.

Four turkeys, young males or jakes, lope for cover along Buffalo Creek. As they disappear beneath spreading cottonwoods, we hear an outpouring from cicadas. Loud as any rivet gun, they call from the stems of sagebrush or fly away accompanied by their own shrill buzzing. The flight appears crude and short, like tiny wind-up airplanes, each exodus presenting a tempting target for hungry nighthawks and kites.

In one of the pastures we find a walkingstick insect, green and maybe two inches long, firmly affixed to the parched stems and leaves of a scurfpea. Even in the devastating heat of August, life abounds on this prairie. And today much of the life literally sings; the air is awash with the scent of rain and clouds are building ever higher as we watch.

It is August 27, the air is hot and stifling, and a walk along Sleeping Bear Creek leaves me drenched with sweat. Yet even though August heat still dominates, you can sense seasonal changes in the early morning cool, the length of daylight, the fact that weather systems have shifted and storms are beginning to reach us once more. September can be either an autumn month or a summer month, but generally it is simply a slow transition out of the August heat and into the pleasant days and cool nights of October.

Days on the Selman Ranch will still reach into the nineties and sometimes top 100, but there will be times when a weather system slips in to cool things off, or bring much-needed rain. Even late August is notorious for a brief cool snap that brings rain to fall on Labor Day activities. But today will be around 92 degrees, the winds out of the southwest, high pressure dominating.

Sleeping Bear Creek is running, the pools full and swarming with minnows. The little fish all face upstream, waiting for food to drift down on the current, and when they maneuver they do so in unison, like a flock of blackbirds on the wing. Brown leopard frogs leap from the bank and into the water seeking underwater shelter. At the same time a few cricket frogs simply leap a foot or so out into the current, then swim back to vegetation along the bank, where they attach themselves to a stem and watch for additional danger.

Those who think August is too hot for a walk haven't seen the tall rushes along

Sleeping Bear lined with pale blue dragonflies with transparent wings. Or the pink flowers on camphor weed, or the first blue asters of autumn. Night prowlers like deer and raccoons have frequented the sand along the stream, and the 'coons have been busy popping open freshwater mussels and feasting on their own brand of cowboy clams. Cattle seek the shade of a big soapberry grove, trampling the soil underneath the trees to fine brown dust. Some of these soapberry trees are exceptionally large, with several two feet in diameter or more. Rocky Mountain bee plant or spiderflower, a big, bushy plant like something you'd expect to purchase in a pot at a shopping mall, brightens up the scenery with purple flowers. In some places the ranch roads are lined with them.

I frighten a great blue heron away from its minnow spearing and remind the turkey vultures circling overhead that I'm not dead yet. These are this year's nestlings trying out their soaring wings, and I don't want them to confuse my lethargy with a need to summon a feathered funeral director. Blue-winged teal loafing and dabbling on a beaver pond casually row away at my approach while cicadas continue their electric hum, something like what insects might sound like if they owned a synthesizer.

A Sleeping Bear Creek tributary cuts deeply through white gypsum and red clay, creating a scenic little canyon complete with a beaver pond. The tributary appears to be spring fed, the water fresh and running. The south-facing gyp outcrop contains a small cave looking out over a mini-forest of cottonwood, sumac, elm, black willow, indigo bush, and rough leaf dogwood. Upon closer examination the beaver pond turns out to be a series of ponds, all ringed with cattails. The north-facing bluff shows an expanse of red clay, and a sizable cottonwood tree is growing straight out of a bank maybe fifteen feet up. A sister tree has been felled by beavers. A small yellow and brown flycatcher darts out from the cottonwood to snare an insect, and at the moment of impact I flush a bobcat out of the grass, which spooks a doe and her fawn from a sumac thicket. I trail all three down to Sleeping Bear Creek just in time to watch a bass leap above a beaver pond and snag a dragonfly out of the air, a play any shortstop would be proud to make. Not bad for an August afternoon, no snakebites and the loss of only a little sweat. Back home in the city, the asphalt parking lots are stifling.

On my way back to the ranch I notice that purple flowers of ironweed are on display atop tall, stout stems. Too many ironweed stalks indicate years of overgrazing or soil disruption, but today the plants are sparse along the little feeder stream flowing into Sleeping Bear Creek. I've been told that the Pawnees used the sap from these stalks as a salve to treat horses with saddle sores. This reminds me that the rancher's curse may be the next miracle cure someday. Indians used hundreds of these prairie plants for medicinal purposes, and today we're just beginning to rediscover some of the curative properties of native plants.

The afternoon seems to belong to the insects: a cicada's recently shucked nymphal skin, clinging to the woody main branch of a clump of sand sagebrush; a red and black eastern milkweed bug, drawing nourishment from the white flowers of wild buckwheat; a pink, black, and white sphinx moth, hovering next to a sand lily, feeding like a hummingbird on the wing.

This insect activity, along with that of meadowlarks and mourning doves, takes place in a wide, flat meadow beside Sleeping Bear Creek. The grasses here are tall and lush, more like the mixtures found in the tallgrass country to the east . . . clumps of Indian grass with golden seed tassels, chest-high tussocks of big bluestem, dense bunches of little bluestem, switchgrass with airy sprays of seeds, knee-high tufts of sideoats grama.

The end of August needs an exclamation point, and mine comes with the discovery of a big western diamondback, about four feet long, coiled in the closely cropped grass along the banks of Buffalo Creek. The snake's head is elevated maybe a foot, pointed like a pistol barrel in my direction, the head waving just a bit, much like those *National Geographic* scenes of a cobra in a basket, entranced by a Hindu flute player.

However, this snake is far from addled. Instead, it is cocked and loaded with magnums, and I ease up until the rattler decides to make a run for cover. In an instant the diamondback is on the move, headed for a plum thicket. And for the first time, the snake begins to rattle, although not much louder than a cicada's buzz.

I take a ranch road back to the bunkhouse, pickup truck windows open, and in a matter of minutes the cab is filled with grasshoppers. I think about Sue's stories of growing up here, and the times when grasshoppers ate the paint off the walls. Or the Christmas Day when her favorite dog was bitten by a rattler. This country has an edge to it, and it can and will bite back. This is the West that Gene and Roy didn't even attempt to immortalize. Maybe that's why you can still see the stars here at night, and the people who hang on here take pride in their independence. They curse the land, yet they love it. Prairies appeal to a peculiar set of people: cowboys, Indians, loners, dreamers. People who need a lot of room and will put up with grasshoppers, rattlesnakes, heat, drought, hail, high winds, and blizzards to make sure they have plenty of space. The cattle are just an excuse, I think, for the opportunity to live the last free life left in America.

The first frost comes early this year, October 6 to be exact. A north wind cuts through summer-weight clothing, and photographer Don House and I search through our gear for coats, a bit amazed by the leaden winter sky and frost on the windshield.

Then just as suddenly as the norther arrived the weather changes moods, the sun returns by midday, the wind dies, and we're standing center stage amid an autumn show of color rivaling the best of spring. As far as one can see are yellow patches of flowering broomweed, scattered sprays of golden asters, gray-green sagebrush heavy with seed, the silver-gray of sagewort, all illuminated by an angle of autumn radiance rivaling the famed light that lures painters to the Mediterranean or New Mexico's Taos. The arching leaves of switchgrass have turned a brilliant gold, while bunches of little bluestem glow a coppery bronze. We find purple gayfeather blooms amid the yellow of broomweed, the color scheme framed by the white flowers of a dense, bushy little aster.

Autumn here can be like a second spring, with daytime temperatures in the seventies and eighties, cool nights, even thunderstorms and tornadoes. In some ways the wildflower show is equal to that of spring. Variety may be lacking, but size and the rich yellows, blues, and purples that prevail are stunning, especially set against the reds, coppers, and golden hues that predominate among the late-season prairie grasses.

With the cooler temperatures comes additional daytime wildlife activity. Northern flickers, the red-shafted or western variety, have moved into the cottonwoods along Buffalo Creek and protest our incursions with calamitous flicker yukking. The green cottonwood leaves now display just a tint of yellow, while brown and yellow leaves invade streamside elms. Soapberry groves are so heavy with fruit that from a distance the trees appear to be a butterscotch color, due to the density of the grape-sized berries that have turned a warm yellow-brown, the color of cinnamon mixed with butter.

We find a plump northern fence lizard sunning on a gypsum ledge above a little beaver pond. The dull gray background color contrasts fine orange lines on the torpid reptile's back. Bluejays are traveling in flocks, noisily announcing the season, while high overhead we hear a wild whinny unlike any other sound in nature. It takes a minute to spot the flock of sandhill cranes circling over the salt flats along the Cimarron.

So to the ethereal music of cranes we inspect the browse line on streamside indigo bush and within minutes flush five whitetail does from a sumac thicket, leaves still green yet streaked with bright crimson. Above the sumac patch is a high bluff with a nice ledge of gypsum, and we decide to use this as an observation post to await the coming of the longhorns. Sue Selman periodically hosts cattle drives on her ranch as a way to celebrate her family heritage. The longhorn herd is owned by one of Sue's neighbors, and the cattle have the best job on the southern plains.

These longhorns are celebrities, available for movies, television, parades, and other public events. Following a traditional chuckwagon breakfast amateur riders take their cue from the local cowboys, who saddle up and commence gathering the longhorns from holding pastures. These drives attract a diverse crowd, from trail riders with a love of western culture to grandmothers with cameras to men who love nothing more than driving a team of horses or mules and a vintage wagon across open country. The cattle know their business and quickly the lead steer takes his position at the point, the others fall in line and away the beautiful animals trot, the mounted entourage in tow eating a little dust in the finest Texas trail drover tradition.

The deer Don and I spooked are still moving slowly down the creek, now at least a mile away. Suddenly they stop, almost sliding to a standstill like trained quarterhorses, then pivot and race for high ground and the safety of the narrow canyons leading up through the gypsum buttes.

Moments later we see what spooked them. The longhorn herd rounds a bend, coming at us in a steady trot, wagons crashing through the sagebrush in their wake. The spectacle is magnificent, and for a minute I feel like I'm witnessing a scene through J.O. Selman's eyes—actually Jimmy Few Clothes, that skinny kid who, threadbare but ambitious, came to the Cimarron in search of the freedom to work hard, dream big, and convert grass to beef. I bet his ghost is watching, too, smiling at the way those longhorns splash across a water crossing and the way domestic cattle in the pasture seem terrified at their approach, as if they, too, had seen a ghost.

The longhorns swing toward the chuckwagon, which is ready for the noon meal, and Don and I have time on our hands to walk the creek, picking up pieces of chert, checking each for signs of knapping. Several good springs issue from these cliffs, each forming clear, cool pools that mirror the surrounding white rock, the big chunks of gypsum that have sloughed away from the rimrock and tumbled down to form dams and barricades.

Below these pools the current slides downhill to flow into even bigger pools formed by beavers. All this beaver engineering has created a diverse wetland system that remains lush in comparison with the dry and dusty uplands. Cattails are thick, as are the forbs that demand year-round moist soil. The broad-tailed rodents, through their incessant drive to convert running water into calm pools, have enriched the beauty and variety of available wildlife habitat and haven't charged a dime for doing their beaver business.

Several aged elms guard a large slab of gypsum crowning a particularly picturesque mesa overlooking the stream, and near the exposed roots of one of the most ancient trees, a very rotund striped skunk is napping. We ease around the critter, giving it ample room to continue dreaming, and amuse ourselves by picking up chunks of crystal, some nearly transparent, also checking a nice outcropping of flint for elusive projectile points.

Minutes later another skunk, this one apparently spooked by the passing longhorns, comes galloping through knee-high prairie grass directly toward me. I stand as still as possible and the animal approaches within 15 feet before detecting some movement or my scent and throwing its tail straight up in the air to let me know its business end is loaded, cocked, and ready.

The skunk bears left and I bear right, and in doing so I flush a fat autumn grasshopper that flies directly into the web of a black and yellow argiope or garden spider, seemingly as big around as a quarter. The web is strung amid stalks of prairie grass, and the grasshopper is quickly ensnared yet continues to kick, strongly yet futilely.

I check my watch and the argiope has its prey wrapped and then totally encased in webbing in less than a minute. Some of these calf-roping cowboys should take lessons.

The day ends the way a perfect afternoon on the prairie should. A bonfire, chuckwagon grub, field crickets chirping a requiem to summer, coyotes calling up the moon. Monarch butterflies settle in for a night's rest on their way to wintering grounds in Mexico. The plains are alive in this season of transition, while our thoughts stand still, savoring the firelight reflections and the coffee. A horse nickers to those hobbled just beyond the chuckwagon, and I can't help but wonder who else might be riding by out there in the dark, unseen yet near in spirit, ready to unsaddle and rest a spell, here in the heart of this good country.

≈≈≈

November is a mixed-up month, with winter waging a tug of war with Indian summer. We have snow and bitter cold early, cold enough to cause the quail hunters to add extra layers of clothing. Then by Thanksgiving and deer season the weather moderates but the strong south winds return, so strong that I have to tie my hat down with a bandana.

The Selman Ranch is rich with game and people flock here for a chance to hunt in this big, dramatic western setting. The quail hunters have come from as far away as New York; the deer hunters are from Florida. Today several nice bucks hang in front of the ranch house, and Sue's son Coli and friends are busy guiding, skinning, preparing trophy heads.

I pick up a rifle and go into the gypsum breaks just north of the bunkhouse. This broken country is covered with dense skunkbrush thickets and it's tough to hunt. Maybe that's why the biggest whitetail buck I've seen in 30 years of looking bursts out of brush 20 yards ahead of me and disappears within two seconds, while I watch with mouth agape and gun dangling.

Later in the day I find a beautiful little meadow out of the wind, tucked away beneath bluffs and bordered by a stream and a dense stand of red cedars. I don't even have time to doze before a doe comes out of the cedar thicket, crosses the clearing a few yards from where I'm sitting, and heads for the high prairie. Just seconds behind her is a young forkhorn buck, sniffing and trailing, his natural caution overcome by lust. This randy youngster stops mid-clearing and takes a long draught of doe scent. It's an easy shot, a chance to make meat. But I remember the huge buck of this morning and decide his age and wisdom make for a better challenge. We'll let this young fellow grow.

The wind lays late in the afternoon and I stop once more to savor the colors. The evening light is intense and the prairies are aglow with it. The little bluestem tussocks are a deep reddish-orange now, the sage a gunmetal gray. Cedars and soapweed yucca remain darkly evergreen, while the light catches the uppermost cottonwood limbs and paints them silver. And the wraparound horizon, famous for its sunsets, stages a spectacular curtain call. A band just above the skyline is bright orange and blue, while puffs of purple clouds drift by, just the way Zane Grey would describe it.

Walking back to the pickup I flush a covey of quail and a cock pheasant, the latter as brightly colored as the sunset. A cottontail rabbit bursts from shelter beneath a clump of sagebrush, and the sounds from a nearby creek bottom indicate a flock of crows are harassing an owl. Soon the tormented one flees its perch in a big cottonwood, the crows streaming after this nightstalking predator that loves a little fillet of Corvid, plucked from a midnight perch.

As the crows vacate the creekbottom and their din diminishes I hear the hammering of flickers. Both the red-shafted western race as well as yellow-shafted eastern birds are active in this narrow, shallow canyon with its sparse forest of cottonwoods and willows. Within a minute's time I note both downy and hairy woodpeckers, along with the ubiquitous meadowlarks and robins.

At last light I notice a pile of feathers on the crest of a hill. It's the remains of a ring-neck duck, far from water. My thoughts drift back to the prairie falcon I saw earlier and I wonder if the winged hunter took his prize from the air, after a long stoop at high speeds, the impact enough to stun the duck or maybe even kill it, the hunter gorging here atop the prairie, the wind rustling its feathers, the view as big as the bird's desire to roam.

We have shared something here today, the falcon and I. Both hunters, the bird more practical and successful, the human already well fed and prone to daydreaming, both beneficiaries of the gift of the land. I have found my place and the falcon his, a place with enough wild nature left in it that a bird can feast and frail human thoughts can flow uninterrupted. That is the magic of these prairie/plains or what is left of them. J.O. Selman sought freedom here and he found it in the natural rhythms of the land, the way wind and rain and grazers interact, the way the countryside offers up its bounty to those with enough patience to let it be. Three generations of caretakers have done just that, and the fact of their stewardship gives me hope. This land, the culture, the heritage, the wildlife, and all the uniquely evolved prairie plants could well survive another generation, and another. They could survive because people here are imbued with a love of place and a sense of history. They, and those like them who sacrifice to keep the old ways and the wild things together, may just represent the West's last best hope. So I'll keep the faith, and offer these words as a token of my gratitude.

Landscape & Vision

Text and Images of the Plains by Don House

Cimarron River and Buffalo Creek, Selman Ranch

Timeless. This word comes to mind so often here, but it's not right, really. Perhaps timeful. It's not so much that time has stood still, but that all times are still present. You can stand on a bluff overlooking the Cimarron River and see no indication of date. And that is the power of the place. You're not standing along a busy interstate trying to read a historic marker; you're looking at history itself, and the crunch of stone off to your right could be a porcupine nosing around or a rancher in his pickup checking his fence line or a Comanche warrior. Nothing in the landscape has changed, really, except for the viewer. Which is why, when I stand over my tripod cursing my awkward bifocals and hear that little noise off in the distance that seems to be getting closer, a jangling of metal perhaps, I find it utterly believable that in a minute or two I will be trying to explain to a troop of mounted cavalry just what the hell I'm doing out here.

Windmill on Wooden Tower, near Selman, Oklahoma

As a boy, I hiked with a group of Boy Scouts for twenty miles through central Texas summer heat, full packs and one-quart canteens, to an abandoned ranch where we would spend the night and find, we were assured, all the water we needed. We stumbled in near dark to find nothing but a windmill, dry and rusted and silent. I will forever hear that screeching of metal when someone reached up and released the vane and the blades began spinning slowly in a breeze we couldn't even feel. Twelve pairs of desperate eyes watched the horizontal pipe, pointing at us like a gun. And just when we had given up and begun to turn away, water as pure and cool as an artesian well gushed out to our waiting lips.

THE
ERMOTOR

Dugout, near the Cimarron River, Selman Ranch

I am someone who craves solitude, who searches it out, who finds the prairie a place of spiritual strength and incredible physical beauty, who finds himself drawn to it over and over. But there is a difference between solitude and solitary confinement, and it's easy to imagine that many men and women who struggled to survive here must have felt as if they'd been given a life sentence.

Mulberry Tree near Selman, Oklahoma

It is a force to be reckoned with, and not just its occasional temper tantrum tornadoes, but the day after day, year after year presence of the prairie wind. It determines what will grow, what will die, what will last, what will disappear, what you will wear, what you will hear. Resistance is possible but it takes a toll. For every harnessing that succeeds, maybe a windmill for water or electricity, there are a dozen failures. Sucked-dry, sandblasted, shriveled-up, blown-away dreams.

Morningstar School, near Selman, Oklahoma

Maybe the most interesting part of this isolated and beautiful little schoolhouse is not in the photograph. My tripod is standing on the roof of an underground storm shelter. After making this exposure, I ventured down into it, carefully, because it looked like the perfect summer retreat for a rattlesnake. There was no graffiti to blur the vision of the scared young faces lined up on the little cement benches on each long side, and the face of the teacher, probably not much older. A teacher who had seen the funnel cloud through the windows as she paced in front of the blackboard, and had controlled her fear and moved the children quickly to the shelter and gave them each responsibilities to keep their minds off the wind: light the candles, count the students, sing a song, read a poem. And who would be first to go up those steps after the wind died, and tell with her eyes, the anxious faces below, that all was well or that the prairie had been swept clean of all their knowledge.

Cottonwood Skeleton, Selman Ranch

The beech trees of the Ozarks and the cottonwoods of the prairie are the most animal-like: something about the skin tones of their bark or the shape of trunk and limbs. Muscled arms and shoulders and thighs. And to come upon a skeleton sprawled in the sage and bluestem is like discovering a killing field, elephants slaughtered for their tusks, bodies left rotting in the prairie sun.

Buffalo Skull, near the Ghost Town of Salt Springs, Oklahoma

These salt springs are such an enigma. I suppose a geologist could explain it to me in terms of rock formations and the underground movement of water, but to me it's miraculous that I can be standing on salt deposits so thick they support my weight and the weight of a tripod and camera, while fresh water flows behind me in the Cimarron and ahead of me in Buffalo Creek. I'm not the first to be drawn to it. Buffalo, native Americans, immigrants, settlers, cowboys, and even corporations of late have felt the pull of the salt springs. I'm content to stand here and wonder if anyone who sees the photograph will believe it's not snow.

Buffalo Wallow, Selman Ranch

I suppose for every generation there are images that burn into our psyches, and one that has become part of mine is a photograph of a buffalo killing field, circa 1870 where hunters slaughtered so many that the bones stretched out like an ocean toward the horizon. The numbers were beyond human comprehension, like the distance to the nearest galaxy. I'm grateful for this wallow. I can stand here and believe that they are still out there, just out of sight over the next ridge, and that if I stand here long enough, they will return and the smell and sound of thousands of bison will envelope and frighten and thrill me.

Bob's Workbench, near Quinlan, Oklahoma

I barely met the man who works here, but by the time this long exposure was finished, I felt as if I'd known him all my life. Everything you need to know about him is somehow right there on that workbench.

'61 Ford, Selman Ranch

She sits next to the bunkhouse with a clear view of the driveway, the corrals, and the main house. I swear she's watching, sad and confused as new pickups and flatbeds fire to cool morning life and begin their chores. Holding a steaming coffee in one hand, I lean gently on her fender and murmur assurances: "Don't worry, they still love you. They just don't want to wear you out. They're saving you for the big jobs."

But she remembers the same look in the eyes of horses and mules nearly half a century ago when she arrived, enveloped in sparkling clean Detroit arrogance.

FORD

Longhorn, Selman Ranch

There are words and phrases that come up in the presence of longhorns that are seldom mentioned around Herefords or Angus or Charlois: majestic, intelligent, resourceful. As I inched closer to this one, I remembered something else a cowhand had mentioned just this morning over coffee: "A person could end up getting their oil checked if they aren't careful."

Sand Plums and Sunset, Selman Ranch

These are the brambles of the Oklahoma prairie. Biologists can speak to their important role to wildlife as food and shelter, but I'm thinking only of jam, tart and red as a cactus flower. The essence of hard-fought survival distilled into a mason jar.

Original Selman Homestead

If you look up the word commitment in the dictionary, there's a picture of someone planting a tree on the Oklahoma prairie.

Spring Valley Cemetery, near Selman, Oklahoma

Nothing is lonelier than a prairie cemetery, nothing. It's not just the starkness of the setting and the lack of trees or decorative landscaping, but the sense of what it was like for the survivors, what they were stumbling back to after they turned away from the grave of someone they loved. Often the stones are solitary or grouped in twos or threes, an individual family's history of love and grief. These are the most powerful, because it seems so impossible to imagine living alone here. And the stones tell a common and bleak story: have a child, child dies, have a child, child dies, have a child, child dies, die young yourself, or disappear. Even today, a hundred years later, when I turn away from photographing a gravestone, I am grateful to see my pickup parked close-by, knowing that I can just drive away.

W
NATHAN C.
WATKINS
DEC 15 1842
MAY 30 1907
AT REST

Cowboy, Selman Ranch Cattle Drive

Has it always been this way? Prairie rancher as jack of all trades: mechanic, farmer, truck driver, preacher, bookkeeper, meteorologist, veterinarian, carpenter, game warden, and cowhand, changing hats and boots to fit the task. Has it always been this way?

Selman Ranch

I know these tracks were made by a pickup, but I prefer to imagine them as wagon ruts and I'm sitting on the hard seat. My wife and children are looking to me for a decision. Do we stay? Do we go on? Did the ones before us survive? Could it possibly get any better than this? Are those storm clouds? .

Bob's Shed, Exterior View #1

The skulls, like the grinding stone, plows, yokes, and pulleys, seem less like collected antiques and more like the owners have parked them there neatly at the end of the workday and will be back refreshed tomorrow morning.

Silos, Salt Springs, Oklahoma

Is there anything that represents optimism more than a silo? The belief in surplus, the need to store excess. Even the shape speaks of swollen, pregnant expectation.

Mule, near Selman, Oklahoma

I said: "What would make a person choose a mule over a horse?"
She said: "People don't realize that it was mostly mules, not horses, that tamed this country and hauled freight and pulled plows and herded cattle. Mules are more sure-footed than horses, they're less likely to step in holes or trip over rocks. They take the heat better, get by on less food, and are less flighty."
I said: "So why doesn't everyone ride mules instead?"
She said: "Well . . . people who like mules . . . they're kind of a different breed themselves, if you know what I mean."

Cottonwood and Sage

When the shutter clicks and the darkslide is replaced, I am going to walk up to this tree. The smell of warm crushed sage will envelop me the whole distance. Since no one else is in sight, I will place my arms as far around the trunk as I can, press my forehead against the bark, and say these words out loud: live long and prosper.

Deer Prints, Salt Flats, Buffalo Creek, Selman Ranch

I don't take deer for granted. I'm old enough to remember when seeing one was an occasion to savor. In my home state of Arkansas, deer were hunted to essential extinction by the early nineteen hundreds, along with bear, turkey, and elk. A little shameful gift from my ancestors. When I stumble onto deer here, on the prairie, I stop and watch. I can see them for a long time as they run; there is a great distance between hiding places. I try to send a silent thank you to them, a wish for a long life, and just as quickly feel shame as an image crosses my mind, another gift from the gene pool from which I drew: I see myself raising the rifle in my arms and setting the sights just ahead of the fleeing animal.

Cimarron River, Selman Ranch

I have more trouble photographing here than anywhere else. It's not a lack of material. In fact, if anything, there is an overwhelming abundance. Like nowhere I've been, the prairie makes the layers clear. There is the bottom, full of flowers, cacti, sage, lichens, rocks, springs, and grasses; a middle layer of gnarled mesquite and massive cottonwoods, bluffs, mesas, and buttes, fences and ruins of homesteads, people, and animals; and the top layer of expansive vistas, rivers moving away to infinity, and uninterrupted sky. So what's the problem? I think it's the loneliness.

Bob's Shed, Exterior View #2

Oklahoma gargoyles. What is this universal attraction to bones? They must speak to us at some primal level. They define things, even when we're not looking for definitions at the moment. Some are secreted away in the plum thickets and grapevine-choked draws, as if embarrassed by their nakedness; others sit high and exposed on bluffs and rises as if proud of their sun-baked sensuality. City folks can argue about evolution versus creationism and theorize about life after death, but prairie ranchers already know the answers.

Fence and Sign, Selman Ranch

Scale is everything. When I focus my lens on an old cedar fence post, or the interesting lines that barbed wire makes as it undulates over the prairie and toward the end of the earth, some practical section of my brain makes automatic calculations: a four-strand barbed-wire fence costs one dollar/foot to build.

In the Ozark mountains that I call home for most of the year, where an eighty-acre farm is common and a cross fence might be one quarter mile long, the decision to separate a pasture to control grazing is a commitment to come up with one thousand dollars. But here, a cross fence could be five miles long and the cost twenty times higher to accomplish the same thing, the decision more painful and difficult to make.

I pull the darkslide from the back of my camera and make an exposure, feeling that ambivalence again: yearning for a life on the plains, grateful I'm not an Oklahoma rancher.

SELMAN RANC
HOLDING BRAND Z
NO
OR
$ 2500 REW

Corral Ruins, Ghost Town of Salt Springs, Oklahoma

You would think the voices would be gone by now. It's been fifty years since a steer walked through this chute, took a last look around, and stepped into a waiting railroad car, but I can still hear them. There are sounds associated with working cattle, with moving and separating animals that don't come many to the ton—sounds that have hidden themselves in the rough-sawn timbers and creosoted posts, waiting for some greenhorn photographer to release them with his touch.

Bob's Shed, Quinlan, Oklahoma

When I stepped into the barn, I stepped back one hundred and fifty years. In fact I stepped back out and reentered several times just for the effect. On one side of the threshold were pickups and tractors and televisions, and on the other were spurs and saddles and bridles and buffalo hide scales and cavalry-issued tools for officers and enlisted men, a general store and repair shop from another era. And in an adjoining building, the equivalent for native Americans: flint knives and hatchets and arrowheads and scrapers, buffalo rib breast plates, pottery and beaded leather bags. A museum director's dream.

But it was the man and his wife that held my attention. They were happy, calm, and serene. Collectors for thirty years, they had long ago discovered the truth of it: the joy is in the finding, not the having. When someone asked why they didn't have locks on all the doors, the couple seemed surprised. They had long ago extracted the real value from each item. What could someone steal?

"Besides," he said, "locks would only keep honest people out."

Sue and Amy Selman, Selman Ranch

This image of Sue Selman and her daughter Amy, an image they don't particularly like, I should add, was taken on my first visit to the ranch on a cold windy day in February. I was struck by these women: the sheer physical strength and beauty of each as individuals, and the force of nature they become together. Mother and daughter.

There is a photographer's instinct to find the definitive shot, the image that sums up the whole damn thing, and this is it. Throw the rest away, the prairie is here in these faces.

Conclusion

I am watching the sun go down on what may be my last visit to the Selman Ranch. There is such an incredible display of color and texture, any photographer worth a damn would be trying to record it, but I'm feeling a little melancholy at the moment, not wanting this evening or this project to come to an end, and decide to leave the camera in its case. The other photographs have been taken and printed, the pastures and rivers and draws and ridges and creeks and cemeteries and fence lines have been walked and rewalked, the lessons have been learned. And *everything* here on the prairie seems to offer a lesson: opportunities taken or lost, choices made, stupid or wise, acts of greed or generosity; the evidence protected by the dry wind for generations to see.

There is a phenomenon that many photographers recognize. I have labeled it *latent image letdown*. It is the clear knowledge that the photograph that you have just taken will never, ever, be as beautiful as the place itself; that trying to convey in two dimensions what is too big even for three, is a kind of arrogance really. And no matter how many compliments you might receive from people who were not there when you took it, you will know it failed. I have felt that inadequacy here on the Oklahoma prairie more than anywhere else I've worked.

What drew me here today was a cattle drive that Sue had organized. Not a reenactment really, no circus-like and demeaning play acting, no gunfights in front of cardboard buildings or any such nonsense, just a little revisiting of her family history. A group of cowhands and visitors drove a herd of a couple hundred Texas longhorns several miles to the ghost town of Salt Springs. A long, hot, dusty, dirty, and dangerous job, and one that has been done on this ranch almost without interruption for the last one hundred and twenty-five years.

I rode in one of the wagons, looking at the world over the backs of a team of horses, watching the cowhands work the cattle, watching the excitement in the faces of the visitors, and watching Sue. She rides comfortably in the saddle and it was clear, even to a greenhorn like me, that this was not her first drive, and she is not an actor. Sue's been on horseback longer than she has been walking, and seeing her today made it easy to imagine the adventure story of her growing up on this ranch, full of horses and mules, dogs and rattlesnakes, quicksand and grass fires and tornadoes, as well as love and family, hired hands and old cowboys. While the rest of us laughed and talked and pestered the cowhands with stupid questions, Sue kept her eyes scanning the prairie, noting the condition of a fence here, the health of native grasses there, always on duty.

I'm wondering if the visitors recognized the choices Sue has made, the care she has given the land. I doubt they knew the names of the prairie plants that were being crunched underfoot as we moved toward the Cimarron, but the wisdom of Sue's choices allowed them to see deer and turkey, prairie chicken and quail, clean water and giant cottonwoods. In an area with a centuries-long legacy of abuse and overuse, she has tried to restore a balance that once was.

I'm also wondering, as the smell of a good chuckwagon dinner drifts over to where I'm sitting, and the sun says its final goodbye, whether I would do as well. As much as I've learned here from Gary and Sue, I've never been tested. I've never had to look out over hundreds of acres of wheat and know that if the crop fails I will lose the farm. I've never had to weigh some short-term effective, but long-term destructive practice against my family's eating for the winter. I've never had to try to cram a few more head of cattle onto an overgrazed pasture to make a mortgage payment. This much I know: doing the right thing takes more work. It's not uncommon to find a federal wildlife biologist or a state extension agent or a retired cowboy in his nineties dropping by Sue's for coffee, and each contributes to the pool of knowledge and history from which Sue draws to make the important decisions that will see her ranch and home into the next century. She reads, she listens, she surfs the net, she researches, she worries, she hopes for the best, she works hard.

Working hard might be the operative phrase. A few years ago I was sitting in a café in Illinois hundreds of miles away from the Selman Ranch and being drawn into a heated conversation at the table next to mine. A group of local farmers were loudly arguing over the meaning of some provision of a recent farm bill. One man in particular was scathing in his denouncement of programs that rewarded the wrong people for doing the wrong things, and forced the right people doing the right things out of business. I worked up the courage to throw a question into the middle of his rantings. I asked him how he dealt with it, what he did when faced with what seemed like total insanity. The table went silent, and he looked at me with an expression that he probably reserved for young children and intellectually challenged adults. "Well," he said, "you work hard and hope you outlast the stupidity."

I often feel a great sadness in the presence of the kind of intense beauty that exists here. Fifty years has taught me the fragility of it. Public domain or private, you leave it knowing that it may never be the same again. All it takes is a new administration in Washington, or an ambitious developer, or lazy local politicians, or plain bad luck and it can be gone in the proverbial blink of an eye. I don't feel that sadness here, now, and it's because of Sue. One person's, one family's commitment to this little piece of the prairie, feels safer, surer than any government promise; Sue's like-minded daughter and sons the best insurance policy. The photographs I've taken, and the words that Gary has written will pertain to the future as well as the past. *We will be able to come back.*

. . . Don House